DESIGN THEORY

C.C. Lindner
C.A. Rodger
Auburn University
Auburn, Alabama

CRC Press
Boca Raton New York

Library of Congress Cataloging-in-Publication Data

Lindner, Charles C.
Design theory / C.C. Lindner and C.A. Rodger.
p. cm. – – (Discrete mathematics and its applications)
Includes bibliographical references and index.
ISBN 0-8493-3986-3 (alk. paper)
1. Combinatorial designs and configurations. I. Rodger, C. A.
(Christopher Andrew) II. Title. III. Series.
QA166.25.L56 1997
511′.6—dc21 97-7965
CIP

Direct all inquiries to CRC Press LLC, 2000 Corporate Blvd., N.W., Boca Raton, Florida 33431.

International Standard Book Number 0-8493-3986-3
Library of Congress Card Number 97-7965
Printed in the United States of America 1 2 3 4 5 6 7 8 9 0
Printed on acid-free paper

The CRC Press Series on

DISCRETE MATHEMATICS AND ITS APPLICATIONS

Series Editor

Kenneth H. Rosen, Ph.D.

AT&T Bell Laboratories

Charles J. Colbourn and Jeffrey H. Dinitz,
The CRC Handbook of Combinatorial Designs

Steven Furino, Ying Miao, and Jianxing Yin,
Frames and Resolvable Designs: Uses, Constructions, and Existence

Daryl D. Harms, Miroslav Kraetzl, Charles J. Colbourn, and John S. Devitt,
Network Reliability: Experiments with A Symbolic Algebra Environment

Alfred J. Menezes, Paul C. van Oorschot, and Scott A. Vanstone,
Handbook of Applied Cryptography

Richard A. Mollin, Quadratics

Douglas R. Stinson, Cryptography: Theory and Practice

Preface

The aim of this book is to teach students some of the most important techniques used for constructing combinatorial designs. To achieve this goal, we focus on several of the most basic designs: Steiner triple systems, latin squares, and finite projective and affine planes. In this setting, we produce these designs of all known sizes, and then start to add additional interesting properties that may be required, such as resolvability or orthogonality. More complicated structures, such as Steiner quadruple systems, are also constructed.

However, we stress that it is the construction techniques that are our main focus. The results are carefully ordered so that the constructions are simple at first, but gradually increase in complexity. Chapter 4 is a good example of this approach: several designs are produced which together eventually produce Kirkman triple systems. But more importantly, not only is the result obtained, but also each design introduced has a construction that contains new ideas, or that reinforces similar ideas developed earlier in a simpler setting. These ideas are then stretched even further when constructing pairs of orthogonal latin squares in Chapter 5.

It is not the intention of this book to give a categorical survey of important results in combinatorial design theory. There are several good books listed in the Bibliography available for this purpose. On completing a course based on this text, students will have seen some fundamental results in the area. Even better, along with this knowledge, they will have at their finger tips a fine mixture of construction techniques, both classic and hot-off-the-press, and it is this knowledge that will enable them to produce many other types of designs not even mentioned here.

Finally, the best feature of this book is its pictures. A precise mathematical description of a construction is not only dry for the students, it is largely incomprehensible! The figures describing the constructions in this text go a long way to helping students understand and enjoy this branch of mathematics, and should be used at ALL opportunities.

Dedication

To my parents Mary and Charles Lindner, my wife Ann, and my sons Tim, Curt, and Jimmy.

To my wife Sue, my daughters Katrina and Rebecca, and my parents Iris and Ian.

Acknowledgements

First and foremost, we are forever indebted to Rosie Torbert for her infinite patience and her superb skills that she used typesetting this book. The book would not exist without her. Thank you!

We would also like to thank the following people who have read through preliminary versions of this book, sometimes used them in their classes, and then forwarded wonderful comments to us: Elizabeth Billington, Italo Dejter, Jeff Dinitz, Hung Lin Fu, Dean Hoffman, Kevin Phelps, Alex Rosa, Doug Stinson, Anne Street, and the design theory students at Auburn University.

About The Authors

Curt Lindner earned a B.S. in mathematics at Presbyterian College and an M.S. and Ph.D. in mathematics from Emory University. After four wonderful years at Coker College he settled at Auburn University in 1969 where he is now Distinguished University Professor of Mathematics.

Chris Rodger is an Alumni Professor of Mathematics at Auburn University. He completed his B.Sc (Hons) with the University Medal and his M.Sc at The University of Sydney, Australia and his Ph.D. at The University of Reading, England before coming to Auburn University in 1982. He was awarded the Hall Medal by The Institute of Combinatorics and its Applications in 1995.

Contents

List of Figures

Chapter 1

Steiner Triple Systems

1.1 The existence problem

A *Steiner triple system* is an ordered pair (S, T), where S is a finite set of *points* or *symbols*, and T is a set of 3-element subsets of S called *triples*, such that each pair of distinct elements of S occurs together in exactly one triple of T. The *order* of a Steiner triple system (S, T) is the size of the set S, denoted by $|S|$.

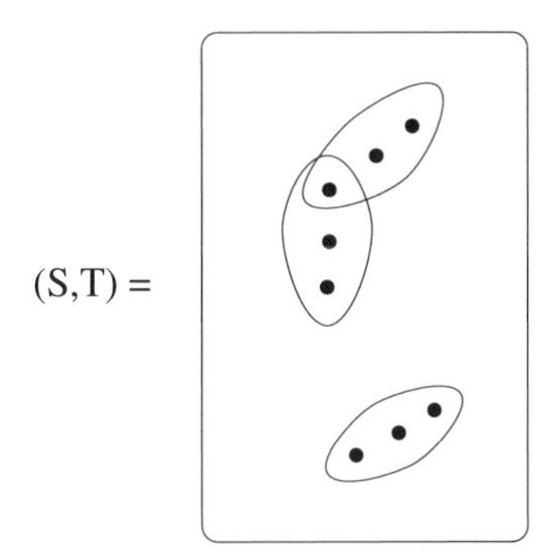

Figure 1.1: *Steiner triple system.*

Example 1.1.1 (a) $S = \{1\}, T = \phi$

(b) $S = \{1, 2, 3\}, T = \{\{1, 2, 3\}\}$

(c) $S = \{1, 2, 3, 4, 5, 6, 7\}, T = \{\{1, 2, 4\}, \{2, 3, 5\}, \{3, 4, 6\}, \{4, 5, 7\}, \{5, 6, 1\}, \{6, 7, 2\}, \{7, 1, 3\}\}$

(d) $S = \{1, 2, 3, 4, 5, 6, 7, 8, 9\}$ and T contains the following triples:

$\{1, 2, 3\}$	$\{1, 4, 7\}$	$\{1, 5, 9\}$	$\{1, 6, 8\}$
$\{4, 5, 6\}$	$\{2, 5, 8\}$	$\{2, 6, 7\}$	$\{2, 4, 9\}$
$\{7, 8, 9\}$	$\{3, 6, 9\}$	$\{3, 4, 8\}$	$\{3, 5, 7\}$

The *complete graph* of order v, denoted by K_v, is the graph with v vertices in which each pair of vertices is joined by an edge. For example, K_7 is shown in Figure 1.2.

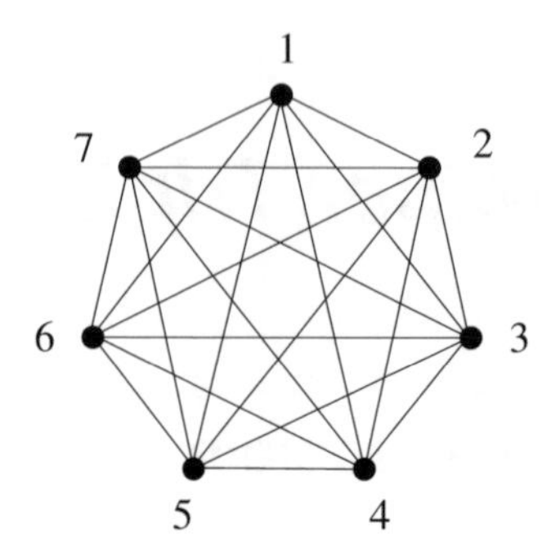

Figure 1.2: *The complete graph* K_7.

A Steiner triple system (S, T) can be represented graphically as follows. Each symbol in S is represented by a vertex, and each triple $\{a, b, c\}$ is represented by a triangle joining the vertices a, b and c. Since each pair of symbols occurs in exactly one triple in T, each edge belongs to exactly one triangle. Therefore a Steiner triple system (S, T) is equivalent to a complete graph $K_{|S|}$ in which the edges have been partitioned into triangles (corresponding to the triples in T).

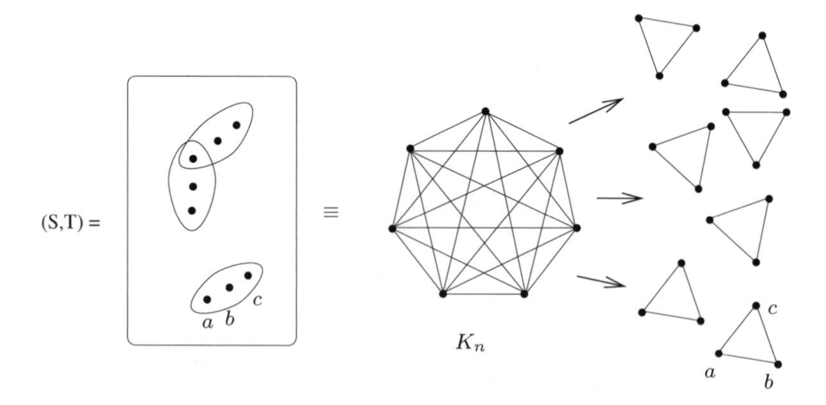

Figure 1.3: *Equivalence between a Steiner triple system and a decomposition of* K_n *into triangles.*

Example 1.1.2 The Steiner triple system of order 7 in Example 1.1.1(c) can be represented graphically by taking the solid triangle joining 1, 2 and 4 below, rotating it once to get the dotted triangle joining $2, 3$ and 5, and continuing this process through 5 more rotations.

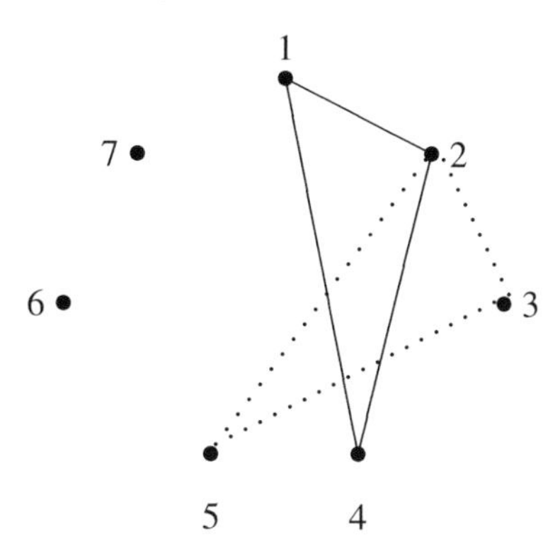

Steiner triple systems were apparently defined for the first time by W. S. B. Woolhouse [28] (Prize question 1733, Lady's and Gentlemens' Diary, 1844) who asked: For which positive integers v does there exist a Steiner triple system of order v? This existence problem of Woolhouse was solved in 1847 by Rev. T. P. Kirkman [12], who proved the following result.

Theorem 1.1.3 *A Steiner triple system of order v exists if and only if $v \equiv 1$ or $3 \pmod 6$.* □

If (S, T) is a triple system of order v, any triple $\{a, b, c\}$ contains the three 2-element subsets $\{a, b\}$, $\{b, c\}$ and $\{a, c\}$, and S contains a total of $\binom{v}{2} = v(v-1)/2$ 2-element subsets. Since every pair of distinct elements of S occurs together in exactly one triple of T, $3|T| = \binom{v}{2}$ and so $|T| = \binom{v}{2}/3$ giving

$$|T| = v(v-1)/6 \tag{1.1}$$

For any $x \in S$, set $T(x) = \{t \backslash \{x\} | x \in t \in T\}$. Then $T(x)$ partitions $S \backslash \{x\}$ into 2-element subsets, and so

$$v - 1 \text{ is even.} \tag{1.2}$$

Since $v-1$ is even, v must be *odd*. A fancy way of saying this is $v \equiv 1, 3$, or $5 \pmod 6$. However, the number of triples $|T| = v(v-1)/6$ is *never* an integer when $v \equiv 5 \pmod 6$, and so we can rule out $v \equiv 5 \pmod 6$ as a possible order of a Steiner triple system. Hence $v \equiv 1$ or $3 \pmod 6$ is a *necessary condition* for the existence of a Steiner triple system of order v.

The next task is to show that for all $v \equiv 1$ or $3 \pmod 6$ there exists a Steiner triple system of order v, which will settle the existence problem for Steiner triple systems. We will give a much simpler proof of this result than the one given by Kirkman. In fact, to demonstrate some of the modern techniques now used in Design Theory, we will actually prove the result several times!

Exercises

1.1.4 Let S be a set of size v and let T be a set of 3-element subsets of S. Furthermore, suppose that

(a) each pair of distinct elements of S belongs to *at least* one triple in T, and

(b) $|T| \leq v(v-1)/6$.

Show that (S,T) is a Steiner triple system.

Remark Exercise 1.1.4 provides a slick technique for proving that an ordered pair (S,T) is a Steiner triple system. It shows that if each pair of symbols in S belongs to at least one triple and if the number of triples is less than or equal to the right number of triples, then each pair of symbols in S belongs to exactly one triple in T.

1.2 $v \equiv 3 \pmod 6$: The Bose Construction

Before presenting the Bose construction [1], we need to develop some "building blocks".

A *latin square* of *order* n is an $n \times n$ array, each cell of which contains exactly one of the symbols in $\{1,2,\ldots,n\}$, such that each row and each column of the array contains each of the symbols in $\{1,2,\ldots,n\}$ exactly once. A *quasigroup* of order n is a pair $(Q,\circ)$, where Q is a set of size n and "$\circ$" is a binary operation on Q such that for every pair of elements $a, b \in Q$, the equations $a \circ x = b$ and $y \circ a = b$ have *unique* solutions. As far as we are concerned a quasigroup is just a latin square with a headline and a sideline.

Example 1.2.1

(a)

1

a latin square of order 1.

$\circ$	1
1	1

a quasigroup of order 1.

(b)

1	2
2	1

a latin square of order 2.

$\circ$	1	2
1	1	2
2	2	1

a quasigroup of order 2.

(c)

1	2	3
3	1	2
2	3	1

a latin square of order 3.

$\circ$	1	2	3
1	1	2	3
2	3	1	2
3	2	3	1

a quasigroup of order 3.

The terms latin square and quasigroup will be used interchangeably.

A latin square is said to be *idempotent* if cell (i, i) contains symbol i for $1 \leq i \leq n$. A latin square is said to be *commutative* if cells (i, j) and (j, i) contain the same symbol, for all $1 \leq i, j \leq n$.

Example 1.2.2 The following latin squares are both idempotent and commutative.

1	3	2
3	2	1
2	1	3

1	4	2	5	3
4	2	5	3	1
2	5	3	1	4
5	3	1	4	2
3	1	4	2	5

The building blocks we need for the Bose construction are idempotent commutative quasigroups of order $2n + 1$.

Exercises

1.2.3 (a) Find an idempotent commutative quasigroup of order

(i) 7,

(ii) 9,
(iii) $2n+1, n \geq 1$. (Hint: Rename the table for $(Z_{2n+1}, +)$, the additive group of integers modulo $2n+1$.)

(b) Show that there are no idempotent commutative latin squares of order $2n, n \geq 1$. (Hint: The symbol 1 occurs in an even number of cells off the main diagonal).

Assuming that idempotent commutative quasigroups of order $2n+1$ exist for $n \geq 1$ (see Exercise 1.2.3), we are now ready to present the Bose Construction [1].

The Bose Construction (for Steiner triple systems of order $v \equiv 3 \pmod 6$).
Let $v = 6n+3$ and let $(Q, \circ)$ be an idempotent commutative quasigroup of order $2n+1$, where $Q = \{1, 2, 3, \ldots, 2n+1\}$. Let $S = Q \times \{1, 2, 3\}$, and define T to contain the following two types of triples.

Type 1: For $1 \leq i \leq 2n+1, \{(i,1), (i,2), (i,3)\} \in T$.

Type 2: For $1 \leq i < j \leq 2n+1, \{(i,1), (j,1), (i \circ j, 2)\}, \{(i,2), (j,2), (i \circ j, 3)\}, \{(i,3), (j,3), (i \circ j, 1)\} \in T$.

Then (S, T) is a Steiner triple system of order $6n+3$.

Type 1 triples.

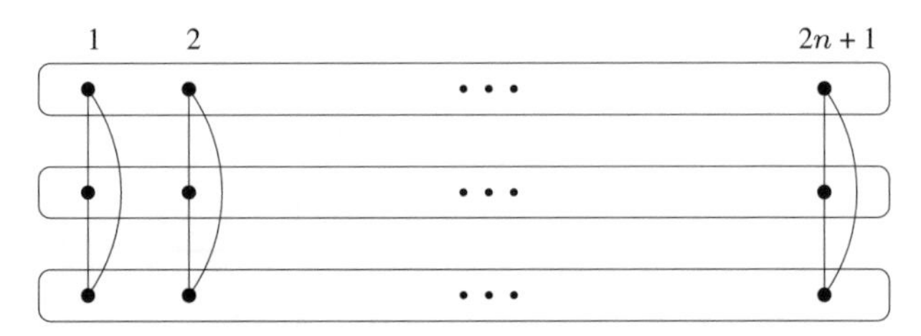

Type 2 triples.

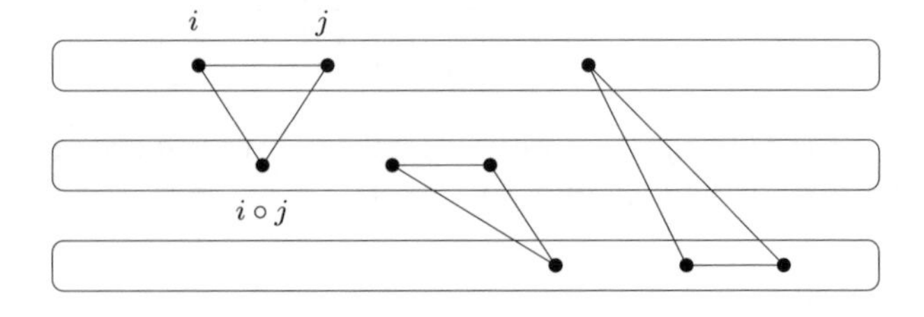

Figure 1.4: *The Bose Construction.*

Before proving this result, it will help tremendously to describe the graphical representation of this construction in Figure 1.4. Since $S = \{1, 2, \ldots, 2n+1\} \times \{1, 2, 3\}$, it makes sense to reflect this structure of S in the graph by drawing the $6n+3$ vertices on 3 "levels" with the $2n+1$ vertices (i, j) on level j, for $1 \leq i \leq 2n+1$ and $1 \leq j \leq 3$.

Proof We prove that (S, T) is a Steiner triple system by using Exercise 1.1.4. We begin by counting the number of triples in T. The number of Type 1 triples is clearly $2n+1$, and in defining the Type 2 triples, there are $\binom{2n+1}{2} = (2n+1)(2n)/2$ choices for i and j, each choice giving rise to 3 triples of Type 2. Therefore

$$\begin{aligned} |T| &= (2n+1) + 3(2n+1)(2n)/2 \\ &= (2n+1)(3n+1) \\ &= v(v-1)/6. \end{aligned}$$

Therefore T contains the right number of triples, and so it remains to show that each pair of distinct symbols in S occurs together in at least one triple of T. Let (a, b) and (c, d) be such a pair of symbols. We consider three cases.

Suppose that $a = c$. Then $\{(a, 1), (a, 2), (a, 3)\}$ is a Type 1 triple in T and contains (a, b) and (c, d).

Suppose that $b = d$. Then $a \neq c$ and so, $\{(a, b), (c, b), (a \circ c, b+1)\} \in T$ and contains (a, b) and (c, d) (of course, the addition in the second coordinate is done modulo 3).

Finally suppose that $a \neq c$ and $b \neq d$. We will assume that $b = 1$ and $d = 2$, as the other cases follow similarly. Since $(Q, \circ)$ is a quasigroup, $a \circ i = c$ for some $i \in Q$. Since $(S, \circ)$ is idempotent and $a \neq c$, it must be that $i \neq a$. Therefore $\{(a, 1), (i, 1), (a \circ i = c, 2)\}$ is a Type 2 triple in T and contains (a, b) and (c, d).

By Exercise 1.1.4 it now follows that (S, T) is a Steiner triple system. □

For convenience, throughout the rest of this text let $STS(v)$ denote a Steiner triple system of order v.

Example 1.2.4 Construct a $STS(9)$ using the Bose construction.

We need an idempotent commutative quasigroup of order $v/3 = 3$, so we will use

$\circ$	1	2	3
1	1	3	2
2	3	2	1
3	2	1	3

.

Let $S = \{1, 2, 3\} \times \{1, 2, 3\}$. Then T contains the following twelve triples:

Type 1: $\{(1,1),(1,2),(1,3)\}, \{(2,1),(2,2),(2,3)\}, \{(3,1),(3,2),(3,3)\}$, and

Type 2:

$i=1, j=2$	$i=1, j=3$
$\{(1,1),(2,1),(1\circ 2=3,2)\}$	$\{(1,1),(3,1),(1\circ 3=2,2)\}$
$\{(1,2),(2,2),(1\circ 2=3,3)\}$	$\{(1,2),(3,2),(1\circ 3=2,3)\}$
$\{(1,3),(2,3),(1\circ 2=3,1)\}$	$\{(1,3),(3,3),(1\circ 3=2,1)\}$
$i=2, j=3$	
$\{(2,1),(3,1),(2\circ 3=1,2)\}$	
$\{(2,2),(3,2),(2\circ 3=1,3)\}$	
$\{(2,3),(3,3),(2\circ 3=1,1)\}$	

Example 1.2.5 The Bose construction and the idempotent commutative quasigroup

(Q, ∘) =

∘	1	2	3	4	5
1	1	5	2	3	4
2	5	2	4	1	3
3	2	4	3	5	1
4	3	1	5	4	2
5	4	3	1	2	5

are used to construct a $STS(v)$. Find:

(a) v (= the order of the triple system).

(b) the triple containing the pair of symbols (a,b) and (c,d) given by:

(i) $(3,1)$ and $(3,3)$

(ii) $(3,2)$ and $(5,2)$

(iii) $(3,2)$ and $(5,3)$.

(a) $(Q,\circ)$ has order $2n+1=5$, so $v=6n+3=15$.

(b) (i) Since $a=c$, the required triple is of Type 1, so is $\{(3,1),(3,2),(3,3)\}$.

(ii) Since the first coordinates are different, the triple is of Type 2. We need to find $3\circ 5=1$ (from $(Q,\circ)$). Then the triple is $\{(3,2),(5,2),(3\circ 5=1,3)\}$.

(iii) Since the first coordinates are different, the triple is of Type 2. Since $3\circ 4=5$ the triple is $\{(3,2),(4,2),(5,3)\}$.

Exercises

1.2.6 Use the Bose construction to find a $STS(v)$, where
(a) $v = 15$ (b) $v = 21$

1.2.7 The Bose construction and

(Q, ∘) =

∘	1	2	3	4	5	6	7
1	1	6	5	7	2	3	4
2	6	2	7	5	4	1	3
3	5	7	3	6	1	4	2
4	7	5	6	4	3	2	1
5	2	4	1	3	5	7	6
6	3	1	4	2	7	6	5
7	4	3	2	1	6	5	7

are used to find a $STS(v)$. Find:

(a) v, and
(b) the triple containing:
(i) $(6, 1)$ and $(7, 2)$
(ii) $(5, 2)$ and $(5, 3)$
(iii) $(3, 1)$ and $(5, 1)$
(iv) $(3, 1)$ and $(5, 3)$
(v) $(4, 3)$ and $(7, 3)$
(vi) $(6, 1)$ and $(2, 2)$
(vii) $(3, 2)$ and $(7, 2)$

1.3 $v \equiv 1$ (mod 6): The Skolem Construction

As in Section 1.2, we will need some building blocks before presenting the Skolem construction [21].

A latin square (quasigroup) L of order $2n$ is said to be *half-idempotent* if for $1 \leq i \leq n$ cells (i, i) and $(n+i, n+i)$ of L contain the symbol i.

Example 1.3.1 The following latin squares (quasigroups) are half-idempotent and commutative.

1	2
2	1

∘	1	2
1	1	2
2	2	1

1	3	2	4
3	2	4	1
2	4	1	3
4	1	3	2

$\circ$	1	2	3	4
1	1	3	2	4
2	3	2	4	1
3	2	4	1	3
4	4	1	3	2

1	4	2	5	3	6
4	2	5	3	6	1
2	5	3	6	1	4
5	3	6	1	4	2
3	6	1	4	2	5
6	1	4	2	5	3

latin squares

$\circ$	1	2	3	4	5	6
1	1	4	2	5	3	6
2	4	2	5	3	6	1
3	2	5	3	6	1	4
4	5	3	6	1	4	2
5	3	6	1	4	2	5
6	6	1	4	2	5	3

quasigroups

Exercises

1.3.2 Find a half-idempotent commutative quasigroup of order

(a) 8

(b) 10

(c) $2n, n \geq 1$. (Hint: Rename the table for $(Z_{2n}, +)$, the additive group of integers modulo $2n$.)

Assuming that half-idempotent commutative quasigroups of order $2n$ exist for all $n \geq 1$ (see Exercise 1.3.2), we are now ready to present the Skolem Construction [21].

The Skolem Construction (Steiner triple systems of order $v \equiv 1 \pmod 6$).
Let $v = 6n+1$ and let $(Q, \circ)$ be a half-idempotent commutative quasigroup of order $2n$, where $Q = \{1, 2, 3, \ldots, 2n\}$. Let $S = \{\infty\} \bigcup (Q \times \{1, 2, 3\})$ and define T as follows:

Type 1: for $1 \leq i \leq n, \{(i, 1), (i, 2), (i, 3)\} \in T$,
Type 2: for $1 \leq i \leq n, \{\infty, (n+i, 1), (i, 2)\}$,
$\{\infty, (n+i, 2), (i, 3)\}, \{\infty, (n+i, 3), (i, 1)\} \in T$, and
Type 3: for $1 \leq i < j \leq 2n, \{(i, 1), (j, 1), (i \circ j, 2)\}$,
$\{(i, 2), (j, 2), (i \circ j, 3)\}, \{(i, 3), (j, 3), (i, \circ j, 1)\} \in T$.
Then (S, T) is a Steiner triple system of order $6n + 1$.

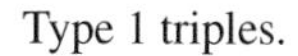

Type 1 triples.

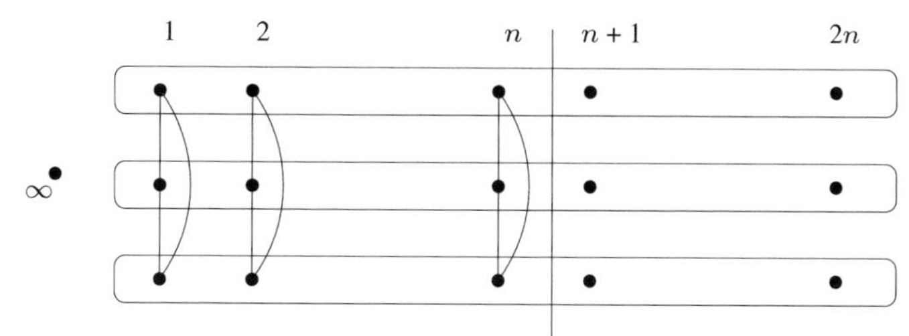

Type 2 triples.

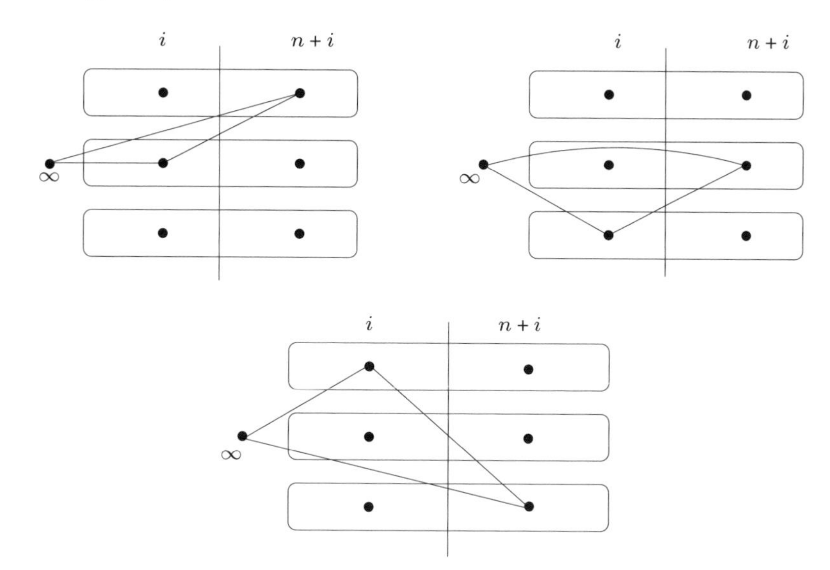

Type 3 triples.

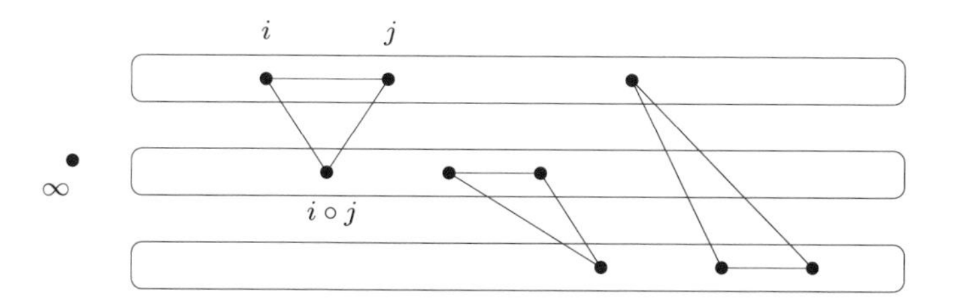

Figure 1.5: *The Skolem Construction.*

Since a picture is worth 1000 definitions, the Skolem Construction is presented graphically in Figure 1.5. The proof that the Skolem Construction actually produces a Steiner triple system is similar to the proof presented in Section 1.2, so is relegated to Exercise 1.3.7.

Example 1.3.3 Construct a $STS(7)$ using the Skolem Construction.

We need a half-idempotent commutative quasigroup of order $(v-1)/3 = 2$, so we use

(Q, ∘) =

∘	1	2
1	1	2
2	2	1

Let $S = \{\infty\} \bigcup (\{1,2\} \times \{1,2,3\})$. Then T contains the following seven triples:

Type 1: $\{(1,1),(1,2),(1,3)\}$,
Type 2: $\{\infty,(2,1),(1,2)\}, \{\infty,(2,2),(1,3)\}, \{\infty,(2,3),(1,1)\}$, and
Type 3: $\{(1,1),(2,1),(1 \circ 2 = 2,2)\}, \{(1,2),(2,2),(1 \circ 2 = 2,3)\}$
$\{(1,3),(2,3),(1 \circ 2 = 2,1)\}$.

Example 1.3.4 The Skolem construction and the half-idempotent commutative quasigroup

(Q, ∘) =

∘	1	2	3	4	5	6
1	1	3	4	2	5	6
2	3	2	5	4	6	1
3	4	5	3	6	1	2
4	2	4	6	1	3	5
5	5	6	1	3	2	4
6	6	1	2	5	4	3

are used to construct a $STS(v)$. Find:

(a) v (= the order of the Steiner triple system)

(b) the triple containing the pair of symbols (a,b) and (c,d), given by:

(i) $(2,1)$ and $(2,2)$
(ii) $(4,1)$ and $(4,2)$
(iii) $(4,1)$ and $(1,2)$
(iv) $(4,2)$ and $(6,2)$.

(a) $(Q,\circ)$ has order $2n = 6$, so $v = 6n+1 = 19$.

(b) (i) Since $a = c = 2 \leq 3 = n$, the required triple is of Type 1, so is $\{(2,1),(2,2),(2,3)\}$.

(ii) Since $a = c = 4 > 3 = n$, the required triple is of Type 3. Since symbol $(4,1)$ occurs one level "above" the symbol $(4,2)$, we solve the equation $4 \circ x = 4$, giving $x = 2$. Therefore the required triple is $\{(a = 4, 1), (x = 2, 1), (a \circ z = 4 = c, 2)\}$

(iii) Since $(4,1)$ occurs one level above $(1,2)$, and since $c+n = 1+3 = 4 = a$, the required triple is of Type 2, and so is $\{\infty, (4,1), (1,2)\}$.

(iv) Since $b = d$, the required triple is of Type 3, so is $\{(a = 4, 2), (c = 6, 2), (a \circ c = 5, 3)\}$.

Exercises

1.3.5 Use the Skolem Construction to find a $STS(v)$, where

(a) $v = 13$

(b) $v = 19$.

1.3.6 The Skolem Construction and

$(Q, \circ) =$

$\circ$	1	2	3	4	5	6	7	8
1	1	3	5	6	2	7	4	8
2	3	2	6	5	4	8	1	7
3	5	6	3	7	8	4	2	1
4	6	5	7	4	3	1	8	2
5	2	4	8	3	1	5	7	6
6	7	8	4	1	5	2	6	3
7	4	1	2	8	7	6	3	5
8	8	7	1	2	6	3	5	4

are used to find a $STS(v)$. Find:

(a) v, and

(b) the triple containing:

(i) $(1,1)$ and $(1,3)$

(ii) $(6,1)$ and $(6,2)$

(iii) $(5,1)$ and ∞

(iv) $(4,1)$ and ∞

(v) $(4,1)$ and $(6,3)$

(vi) $(7,1)$ and $(7,3)$

(vii) $(4,1)$ and $(5,1)$

(viii) $(2,3)$ and $(6,2)$

(ix) $(5, 2)$ and $(7, 2)$
(x) $(1, 3)$ and $(7, 1)$

1.3.7 (a) In the Skolem Construction, count the number of triples of Types 1, 2 and 3, and show that this number is $v(v-1)/6$ (where v is the order of the STS).

(b) Prove that the Skolem Construction produces a $STS(6n+1)$, by using part (a) and Exercise 1.1.4.

1.4 $v \equiv 5 \pmod 6$: The $6n+5$ Construction

We have managed to construct Steiner triple systems of all orders $\equiv 1$ or $3 \pmod 6$. However no $STS(6n+5)$ exists. But we can get very close!

At this point it becomes necessary to generalize Steiner triple systems. A *pairwise balanced design* (or simply, PBD) is an ordered pair (S, B), where S is a finite set of symbols, and B is a collection of subsets of S called *blocks*, such that each pair of distinct elements of S occurs together in exactly one block of B. As with triple systems $|S|$ is called the *order* of the PBD. So a STS is a pairwise balanced design in which each block has size 3.

Our immediate need (to be used in Section 1.5) is to produce a PBD (S, B) of order v with exactly one block of size 5 and the rest having size 3, for all $v \equiv 5 \pmod 6$. So only the one block of size 5 stops this PBD from being a STS!

Example 1.4.1 (a) $S = \{1, 2, 3, 4, 5\}, B = \{\{1, 2, 3, 4, 5\}\}$

(b) $S = \{1, 2, \ldots, 11\}$ and B contains the following blocks:

$\{1, 2, 3, 4, 5\}$	$\{2, 6, 9\}$	$\{3, 7, 8\}$	$\{4, 8, 11\}$
$\{1, 6, 7\}$	$\{2, 7, 11\}$	$\{3, 9, 10\}$	$\{5, 6, 8\}$
$\{1, 8, 9\}$	$\{2, 8, 10\}$	$\{4, 6, 10\}$	$\{5, 7, 10\}$
$\{1, 10, 11\}$	$\{3, 6, 11\}$	$\{4, 7, 9\}$	$\{5, 9, 11\}$

Exercises

1.4.2 Equation (1.1) establishes that in a $STS(v)$ there are $v(v-1)/6$ triples. Letting $v = 6n+5$, find the number of triples (blocks of size 3) in a PBD of order v with one block of size 5, the rest of size 3.

The following construction is a modification of the Bose Construction, so we already have the relevant building block, namely idempotent commutative quasigroups of order $2n+1$.

The $6n+5$ Construction. Let $(Q, \circ)$ be an idempotent commutative quasigroup of order $2n+1$, where $Q = \{1, 2, \ldots, 2n+1\}$, and let α be the permutation $(1)(2, 3, 4, \ldots, 2n+1)$. Let $S = \{\infty_1, \infty_2\} \bigcup (\{1, 2, \ldots, 2n+1\} \times$

$\{1, 2, 3\}$) and let B contain the following blocks:

Type 1: $\{\infty_1, \infty_2, (1, 1), (1, 2), (1, 3)\}$,
Type 2: $\{\infty_1, (2i, 1), (2i, 2)\}, \{\infty_1, (2i, 3), ((2i)\alpha, 1)\}$,
$\{\infty_1, ((2i)\alpha, 2), ((2i)\alpha, 3)\}, \{\infty_2, (2i, 2), (2i, 3)\}$
$\{\infty_2, ((2i)\alpha, 1), ((2i)\alpha, 2)\}, \{\infty_2, (2i, 1), ((2i)\alpha^{-1}, 3)\}$
for $1 \le i \le n$, and
Type 3: $\{(i, 1), (j, 1), (i \circ j, 2)\}, \{(i, 2), (j, 2), (i \circ j, 3)\}, \{(i, 3), (j, 3)((i \circ j)\alpha, 1)\}$, for $1 \le i < j \le 2n + 1$.

Then (S, B) is a $PBD(6n + 5)$ with exactly one block of size 5 and the rest of size 3.

Type 1 block and Type 2 triples.

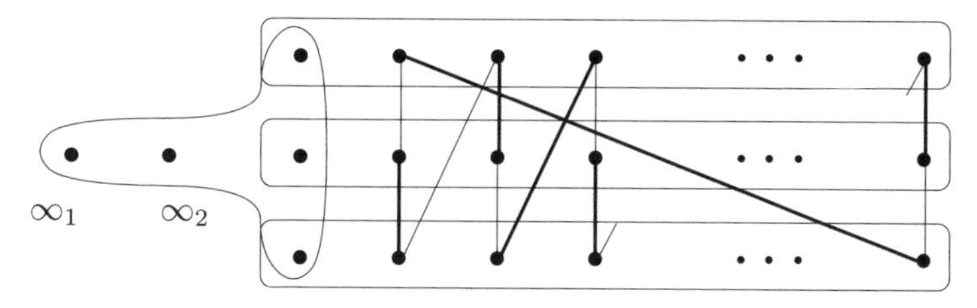

Type 3 triples.

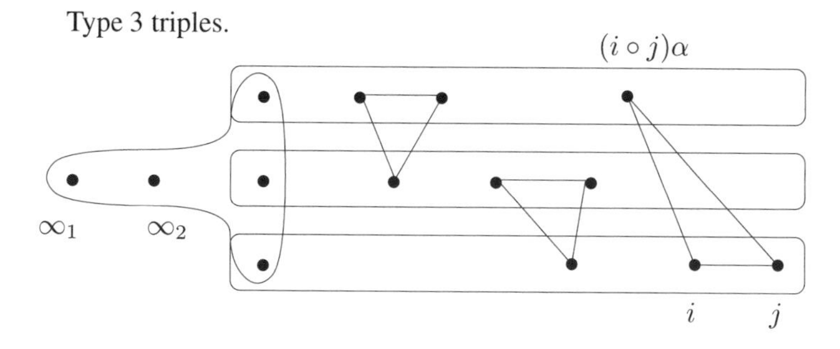

Figure 1.6: *The $6n + 5$ Construction.*

See Figure 1.6 for a graphical representation of this construction. In Figure 1.6, the thin lines in the cycle of length $6n$ join symbols that occur in a triple with ∞_1, the heavy lines in this cycle join symbols that occur in a triple with ∞_2; these are the Type 2 triples.

Example 1.4.3 Use the $6n+5$ construction to find a PBD of order $6n+5 = 11$ with one block of size 5 and the rest of size 3.

We begin with the idempotent commutative quasigroup $(Q, \circ)$ of order $2n+1 = 3$:

$\circ$	1	2	3
1	1	3	2
2	3	2	1
3	2	1	3

$(Q, \circ) =$ (the table above)

Define $\alpha = (1)(2,3)$, so $1\alpha = 1, 2\alpha = 3$ and $3\alpha = 2$. Then $S = \{\infty_1, \infty_2\} \cup (\{1,2,3\} \times \{1,2,3\})$, and B contains the following blocks:

Type 1: $\{\infty_1, \infty_2, (1,1), (1,2), (1,3)\}$

Type 2: $\{\infty_1, (2,1), (2,2)\}, \{\infty_1, (2,3), (3,1)\}, \{\infty_1, (3,2), (3,3)\}$, $\{\infty_2, (2,2), (2,3)\}, \{\infty_2, (3,1), (3,2)\}, \{\infty_2, (3,3), (2,1)\}$, and

Type 3:

$i = 1, j = 2$

$\{(1,1), (2,1), (1 \circ 2 = 3, 2)\}$
$\{(1,2), (2,2), (1 \circ 2 = 3, 3)\}$
$\{(1,3), (2,3), ((1 \circ 2)\alpha = 3\alpha = 2, 1)\}$

$i = 1, j = 3$

$\{(1,1), (3,1), (1 \circ 3 = 2, 2)\}$
$\{(1,2), (3,2), (1 \circ 3 = 2, 3)\}$
$\{(1,3), (3,3), ((1 \circ 3)\alpha = 2\alpha = 3, 1)\}$

$i = 2, j = 3$

$\{(2,1), (3,1), (2 \circ 3 = 1, 2)\}$
$\{(2,2), (3,2), (2 \circ 3 = 1, 3)\}$
$\{(2,3), (3,3), ((2 \circ 3)\alpha = 1\alpha = 1, 1)\}$

Exercises

1.4.4 Use the $6n+5$ Construction to find a PBD of order 17 with one block of size 5, the rest of size 3.

1.4.5 A $PBD(17)$ with one block of size 5, the rest of size 3 is constructed using the $6n + 5$ Construction and the following idempotent commutative quasigroup.

$(Q, \circ) =$

$\circ$	1	2	3	4	5
1	1	5	2	3	4
2	5	2	4	1	3
3	2	4	3	5	1
4	3	1	5	4	2
5	4	3	1	2	5

Find the blocks containing the following pairs:

(i) ∞_1 and $(3, 2)$
(ii) $(3, 3)$ and $(5, 3)$
(iii) $(4, 3)$ and $(5, 1)$
(iv) $(1, 1)$ and $(1, 3)$
(v) $(3, 3)$ and $(4, 3)$
(vi) $(3, 1)$ and $(5, 3)$
(vii) $(2, 1)$ and $(5, 3)$
(viii) $(1, 1)$ and $(5, 3)$

1.4.6 Let (P, B) be a pair where P is a set of size n and B is a collection of subsets (blocks) of P. Suppose that

(a) each pair of distinct elements of P belongs to a block of B, and
(b) the blocks of B cover $\leq \binom{n}{2}$ 2-element subsets of P.

Prove that (P, B) is a PBD.

1.4.7 Use the proof techniques in Exercise 1.4.6 to prove that the $6n + 5$ Construction does indeed give a PBD (with one block of size 5 and the remaining blocks of size 3).

1.5 Quasigroups with holes and Steiner triple systems

1.5.1 Constructing quasigroups with holes

Let $Q = \{1, 2, \ldots, 2n\}$ and let $H = \{\{1, 2\}, \{3, 4\}, \ldots, \{2n - 1, 2n\}\}$. In what follows, the two element subsets $\{2i - 1, 2i\} \in H$ are called *holes*. A *quasigroup with holes* H is a quasigroup $(Q, \circ)$ of order $2n$ in which for each $h \in H$, $(h, \circ)$ is a subquasigroup of $(Q, \circ)$. The following example makes this clear.

Example 1.5.1 For each $h \in H$ we need to choose a subquasigroup $(h, \circ)$. For each $h = \{x, y\} \in H$, we have two choices:

$\circ$	x	y
x	x	y
y	y	x

or

$\circ$	x	y
x	y	x
y	x	y

The following are quasigroups with holes H of orders 6 and 8. The subquasigroup in the cells $h \times h$ are bordered by heavy lines and the symbols they contain are in bold type. These examples are also commutative.

$\circ$	1	2	3	4	5	6
1	**1**	**2**	5	6	3	4
2	**2**	**1**	6	5	4	3
3	5	6	**3**	**4**	1	2
4	6	5	**4**	**3**	2	1
5	3	4	1	2	**5**	**6**
6	4	3	2	1	**6**	**5**

$\circ$	1	2	3	4	5	6	7	8
1	**1**	**2**	5	6	7	8	3	4
2	**2**	**1**	8	7	3	4	6	5
3	5	8	**3**	**4**	2	7	1	6
4	6	7	**4**	**3**	8	1	5	2
5	7	3	2	8	**5**	**6**	4	1
6	8	4	7	1	**6**	**5**	2	3
7	3	6	1	5	4	2	**7**	**8**
8	4	5	6	2	1	3	**8**	**7**

We now solve the existence problem for commutative quasigroups with holes H. It turns out that they are easy to construct if their order is 2 (mod 4) using a direct product (see Exercise 1.5.10), but to construct one of every even order we will need the following more complicated construction given in Theorem 1.5.5.

In this construction, it will be necessary to rename the symbols of a $PBD(v)$, $v \neq 5$ with at most one block of size 5 and the rest of size 3, with the symbols in $\{1, 2, \ldots, v\}$ so that it contains the triples $\{1, 2, v\}$, $\{3, 4, v\}, \ldots,$ $\{v-2, v-1, v\}$. It is easy to see that this is always possible, but the following example may also help.

Example 1.5.2 Consider the $STS(9)$ constructed in Example 1.2.4. We can arbitrarily choose any symbol to be renamed $v = 9$, say symbol $(1, 1)$. We can also arbitrarily pick another symbol to be renamed 1, say symbol $(1, 2)$. This determines symbol 2, because we want $\{1, 2, 9\}$ to be a triple and we know that our $STS(9)$ contains the triple $\{(1, 1), (1, 2), (1, 3)\}$; so $(1, 3)$ is renamed 2. Similarly, if we rename $(2, 1)$ with 3 (again, arbitrarily chosen), then since $\{(1, 1),$ $(2, 1), (3, 2)\}$ is a triple we must rename $(3, 2)$ with 4 in order that $\{3, 4, 9\}$ is a triple. Completing this process: we could rename $(3, 1)$ with 5, so $(2, 2)$ becomes 6; and $(2, 3)$ with 7, so $(3, 3)$ becomes 8.

Exercises

1.5.3 Rename the symbols in the $STS(7)$ constructed in Example 1.3.3 with the symbols in $\{1, 2, \ldots, 7\}$ so that $\{1, 2, 7\}$, $\{3, 4, 7\}$ and $\{5, 6, 7\}$ are triples.

1.5.4 Rename the symbols in the $PBD(11)$ constructed in Example 1.4.1 with the symbols $\{1, 2, \ldots, 11\}$ so that the resulting PBD contains the triples $\{1, 2, 11\}$, $\{3, 4, 11\}$, $\{5, 6, 11\}$, $\{7, 8, 11\}$ and $\{9, 10, 11\}$. (Hint: The symbol chosen to be $v = 11$ is not quite an arbitrary choice here, since clearly it cannot be a symbol that occurs in the block of size 5.)

Theorem 1.5.5 *For all $n \geq 3$ there exists a commutative quasigroup of order $2n$ with holes $H = \{\{1, 2\}, \{3, 4\}, \ldots, \{2n-1, 2n\}\}$.*

Proof Let $S = \{1, 2, \ldots, 2n+1\}$. If $2n + 1 \equiv 1$ or $3 \pmod 6$ then let (S, B) be a Steiner triple system of order $2n+1$ (see Sections 1.2 and 1.3), and if $2n+1 \equiv 5 \pmod 6$ then let (S, B) be a PBD of order $2n+1$ with exactly one block, say b, of size 5, and the rest of size 3 (see Section 1.4). By renaming the symbols in the triples (blocks) if necessary, we can assume that the only triples containing symbol $2n+1$ are:

$$\{1, 2, 2n+1\}, \{3, 4, 2n+1\}, \ldots, \{2n-1, 2n, 2n+1\}.$$

(In forming the quasigroup, these triples are ignored.)

Define a quasigroup $(Q, \circ) = (\{1, 2, \ldots, 2n\}, \circ)$ as follows:

(a) For each $h \in H = \{\{1, 2\}, \{3, 4\}, \ldots, \{2n-1, 2n\}\}$ let $(h, \circ)$ be a subquasigroup of $(Q, \circ)$;

(b) for $1 \leq i \neq j \leq 2n$, $\{i, j\} \notin H$ and $\{i, j\} \not\subseteq b$, let $\{i, j, k\}$ be the triple in B containing symbols i and j and define $i \circ j = k = j \circ i$; and

(c) if $2n + 1 \equiv 5 \pmod 6$ then let $(b, \otimes)$ be an idempotent commutative quasigroup of order 5 (see Example 1.2.2) and for each $\{i, j\} \subseteq b$ define $i \circ j = i \otimes j = j \circ i$.

□

Example 1.5.6 Construct a commutative quasigroup with holes H of order 8 and one of order 10.

To construct such a quasigroup of order $2n = 8$, we use the following STS of order $2n + 1 = 9$:

1	2	9	3	6	7	4	5	8
3	4	9	1	6	8	2	5	7
5	6	9	1	4	7	2	3	8
7	8	9	1	3	5	2	4	6

(This $STS(9)$ can be found by renaming the symbols $(1,1), (1,2), (1,3), (2,1), (2,2), (2,3), (3,1), (3,2)$ and $(3,3)$ of the $STS(9)$ in Example 1.2.4 with the symbols $9, 1, 2, 3, 6, 7, 5, 4$ and 8 respectively (see Example 1.5.2). This apparently strange renaming of the symbols is chosen so that $\{1,2,9\}, \{3,4,9\}, \{5,6,9\}$ and $\{7,8,9\}$ are triples, as is required by the construction in the proof of Theorem 1.5.5.) We now ignore all triples containing symbol 9, and for each other triple, such as $\{3,6,7\}$, define

$$3 \circ 6 = 7 = 6 \circ 3, \quad 3 \circ 7 = 6 = 7 \circ 3, \text{ and } 6 \circ 7 = 3 = 7 \circ 6$$

to produce the following partial quasigroup.

$\circ$	1	2	3	4	5	6	7	8
1			5	7	3	8	4	6
2			8	6	7	4	5	3
3	5	8			1	7	6	2
4	7	6			8	2	1	5
5	3	7	1	8			2	4
6	8	4	7	2			3	1
7	4	5	6	1	2	3		
8	6	3	2	5	4	1		

Adding the appropriate symbols in the cells in $h \times h, h \in H$ (see (a) in the proof of Theorem 1.5.5) gives the required quasigroup.

To construct a commutative quasigroup with holes of order $2n = 10$, we use the following PBD of order $2n + 1 = 11$ with exactly one block of size 5, the rest of size 3:
$\{1, 2, 11\}$, $\{3, 4, 11\}$, $\{5, 6, 11\}$, $\{7, 8, 11\}$, $\{9, 10, 11\}$, $\{1, 4, 7\}$, $\{1, 6, 10\}$, $\{2, 3, 6\}$, $\{2, 4, 9\}$, $\{2, 5, 7\}$, $\{2, 8, 10\}$, $\{3, 7, 10\}$, $\{4, 5, 10\}$, $\{4, 6, 8\}$, $\{6, 7, 9\}$, $\{1, 3, 5, 8, 9\}$.
(This $PBD(11)$ can be produced from the $PBD(11)$ in Example 1.4.1 by renaming the symbols 1, 2, 3, 4, 5, 6, 7, 8, 9, 10, and 11 with 1, 8, 3, 9, 5, 4, 7, 10, 6, 2, and 11 respectively. Again, this renaming of the symbols is chosen so that $\{1, 2, 11\}, \{3, 4, 11\}, \ldots, \{9, 10, 11\}$ are triples, as required.) Ignoring all triples containing symbol 11, and using the following quasigroup $(\{1, 3, 5, 8, 9\}, \otimes)$ of order 5 to define $a \circ b$ for each pair of symbols in the block $\{1, 3, 5, 8, 9\}$ of size 5,

$\otimes$	1	3	5	8	9
1	1	8	3	9	5
3	8	3	9	5	1
5	3	9	5	1	8
8	9	5	1	8	3
9	5	1	8	3	9

we obtain the following partial quasigroup.

$\circ$	1	2	3	4	5	6	7	8	9	10
1			**8**	7	**3**	10	4	**9**	**5**	6
2			6	9	7	3	5	10	4	8
3	**8**	6			**9**	2	10	**5**	1	7
4	7	9			10	8	1	6	2	5
5	**3**	7	**9**	10			2	**1**	**8**	4
6	10	3	2	8			9	4	7	1
7	4	5	10	1	2	9			6	3
8	**9**	10	**5**	6	**1**	4			**3**	2
9	**5**	4	**1**	2	**8**	7	6	**3**		
10	6	8	7	5	4	1	3	2		

Adding the appropriate symbols in the cells in $h \times h, h \in H$ (see (a) in the proof of Theorem 1.5.5) produces the required quasigroup. (In the above quasigroup, the products defined by the block of size 5 are in bold type.)

Exercises

1.5.7 Use the $STS(13)$ constructed in Exercise 1.3.5(a) and follow Example 1.5.6 to produce a commutative quasigroup with holes H of order 12.

1.5.8 Use the $PBD(11)$ constructed in Example 1.4.3 and follow Example 1.5.6 to produce a commutative quasigroup with holes H of order 10.

1.5.9 Use the $PBD(17)$ constructed in Exercise 1.4.4 and follow Example 1.5.6 to produce a commutative quasigroup with holes H of order 16.

1.5.10 Let $(\{1,2\}, \circ_1)$ be any quasigroup of order 2 (there are just two; see Example 1.5.1), $(Q, \circ_2)$ an idempotent commutative quasigroup of order $2n+1$, and set $S = \{1,2\} \times Q$. Define a binary operation "$\otimes$" on S by: $(a,c) \otimes (b,d) = (a \circ_1 b, c \circ_2 d)$. Then $(S, \otimes)$ is a commutative quasigroup of order $4n+2$ with holes $H = \{\{(1,i),(2,i)\} \mid i \in Q\}$. (Of course, the above definition is just the *direct product* of quasigroups, which we study in great detail in Chapter 5.)

Starting with the idempotent commutative quasigroup of order 3, construct a commutative quasigroup of order 6 with holes. Then rename the symbols with $1,2,3,4,5,6$ so that the holes are $\{1,2\}$, $\{3,4\}$, and $\{5,6\}$.

1.5.2 Constructing Steiner triple systems using quasigroups with holes

Now that we know that commutative quasigroups with holes H of order $2n$ exist for all $n \geq 3$, we can use these in the next construction. This construction uses commutative quasigroups of order $2n$ with holes H to produce a $STS(6n+1)$, a $STS(6n+3)$ and a $PBD(6n+5)$. This construction is depicted graphically in Figure 1.7.

The Quasigroup with Holes Construction.

Let $(\{1,2,\ldots,2n\}, \circ)$ be a commutative quasigroup of order $2n$ with holes H. Then

(a) $(\{\infty\} \cup (\{1,2,\ldots,2n\} \times \{1,2,3\}), B)$ is a $STS(6n+1)$, where B is defined by:

(1) for $1 \leq i \leq n$ let B_i contain the triples in a $STS(7)$ on the symbols $\{\infty\} \cup (\{2i-1, 2i\} \times \{1,2,3\})$ and let $B_i \subseteq B$, and

(2) for $1 \leq i \neq j \leq 2n$, $\{i,j\} \notin H$ place the triples $\{(i,1),(j,1),(i \circ j,2)\}$, $\{(i,2),(j,2),(i \circ j,3)\}$ and $\{(i,3),(j,3),(i \circ j,1)\}$ in B,

(b) $(\{\infty_1, \infty_2, \infty_3\} \cup (\{1, 2, \ldots, 2n\} \times \{1, 2, 3\}), B')$ is a $STS(6n+3)$, where B' is defined by replacing (1) in (a) with:

(1′) for $1 \le i \le n$ let B'_i contain the triples in a $STS(9)$ on the symbols $\{\infty_1, \infty_2, \infty_3\} \cup (\{2i-1, 2i\} \times \{1, 2, 3\})$ in which $\{\infty_1, \infty_2, \infty_3\}$ is a triple, and let $B'_i \subseteq B'$, and

(c) $(\{\infty_1, \infty_2, \infty_3, \infty_4, \infty_5\} \cup (\{1, 2, \ldots, 2n\} \times \{1, 2, 3\}), B'')$ is a $PBD(6n+5)$ with one block of size 5, the rest of size 3, where B'' is defined by replacing (1) in (a) with:

(1″) for $1 \le i \le n$ let B''_i contain the blocks in a $PBD(11)$ on the symbols $\{\infty_1, \infty_2, \infty_3, \infty_4, \infty_5\} \cup (\{2i-1, 2i\} \times \{1, 2, 3\})$ in which $\{\infty_1, \infty_2, \infty_3, \infty_4, \infty_5\}$ is a block, and let $B''_i \subseteq B''$.

It is straight forward to see that (a) and (b) are triple systems, whereas (c) is a PBD with one block of size 5 and the remaining blocks of size 3. (See Exercises 1.1.4 and 1.4.6). For a pictorial representation of this construction, see Figure 1.7.

Example 1.5.11 Suppose we wish to construct a $STS(31)$ using the Quasigroup with Holes Construction. Since $31 \equiv 1 \pmod 6$, we use (a) in this construction. In step (1) we need to define B_i, so let B_i contain the triples (each B_i can be any $STS(7)$ defined on the symbols $\{\infty\} \cup (\{2i-1, 2i\} \times \{1, 2, 3\})$:

{∞, (2i-1, 1), (2i,1)}	{(2i-1, 1), (2i-1, 2), (2i-1, 3)}
{∞, (2i -1, 2), (2i, 2)}	{(2i-1, 1), (2i, 2), (2i, 3)}
{∞, (2i-1, 3), (2i, 3)}	{(2i, 1), (2i-1, 2), (2i, 3)}
	{(2i, 1), (2i, 2), (2i-1, 3)}

In step (2) we need a commutative quasigroup with holes H of order $2n = 10$, so we can use the one defined in Example 1.5.6. Since v is so large, rather than find the entire $STS(31)$, let's find the triple containing the pair of symbols (i) $(4, 1)$ and $(5, 2)$ (ii) $(5, 1)$ and $(6, 2)$ (iii) ∞ and $(8, 3)$.

(i) Since $\{4, 5\} \notin H$, step 2 of the Quasigroup with Holes Construction shows that we proceed to find the triple as in the Bose Construction. So the triple is $\{(4, 1), (10, 1), (5, 2)\}$ since in the quasigroup $4 \circ 10 = 5$.

(ii) Since $\{5, 6\} \in H$, step 1 of the Quasigroup with Holes Construction shows that we find the triple in B_3. Now, B_i contains the triple $\{(2i-1, 1), (2i, 2), (2i, 3)\}$ so B_3 contains the triple $\{(5, 1), (6, 2), (6, 3)\}$.

(iii) Any triple containing ∞ is defined in step 1 of the Quasigroup with Holes Construction. B_i contains the triple $\{\infty, (2i-1, 3), (2i, 3)\}$, so B_4 contains the triple $\{\infty, (7, 3), (8, 3)\}$.

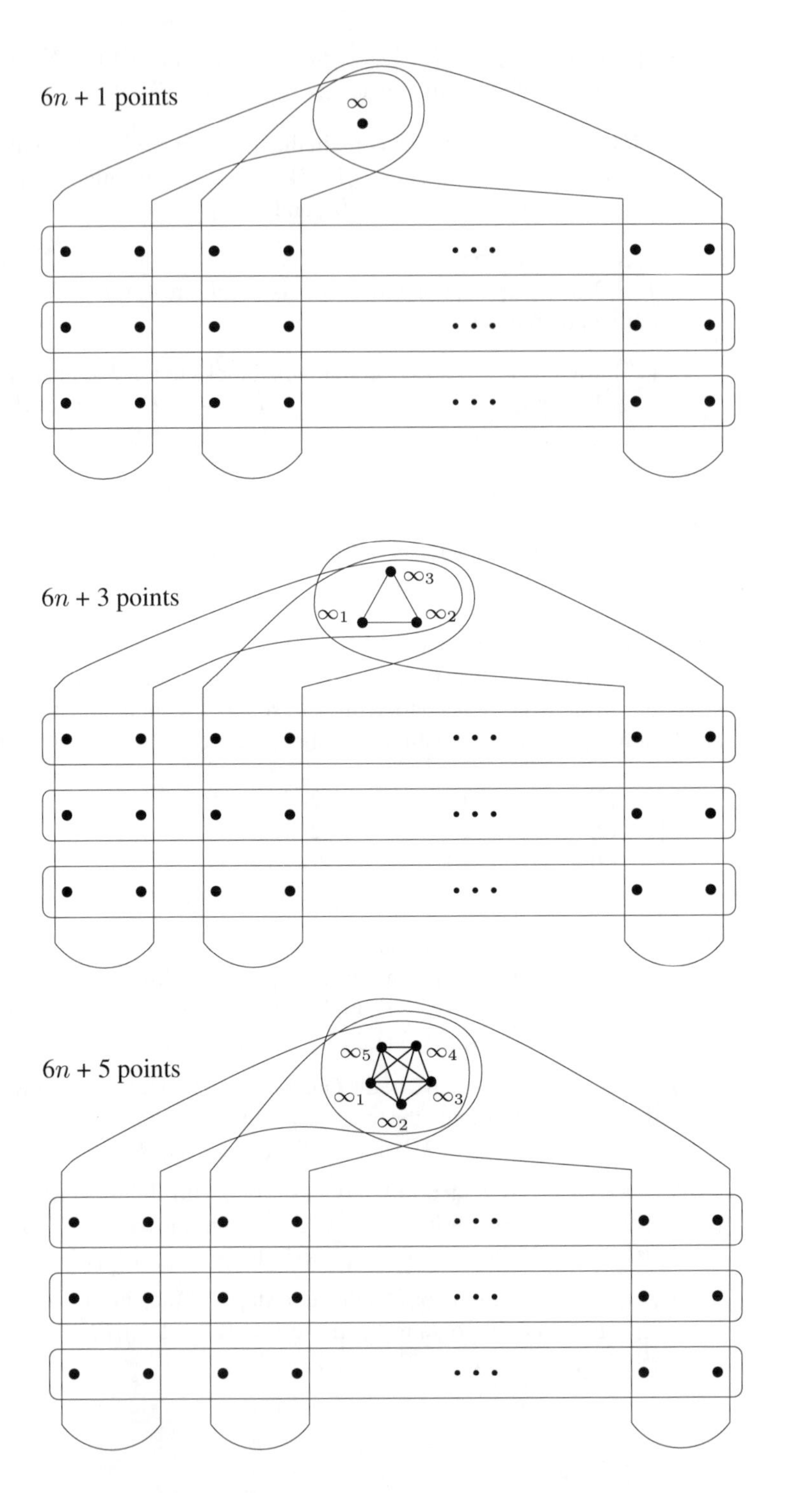

Figure 1.7: *The Quasigroup with Holes Construction.*

Example 1.5.12 Suppose we construct a $STS(33)$ using the Quasigroup with Holes Construction. Since $33 \equiv 3 \pmod 6$ we use (b) in this construction. In step (1) we need to define B_i', so let B_i' contain the following triples (each B_i' can be any $STS(9)$ defined on the symbols $\{\infty_1, \infty_2, \infty_3\} \cup (\{2i-1, 2i\} \times \{1,2,3\})$ providing the symbols are named so that $\{\infty_1, \infty_2, \infty_3\}$ is a triple):

$\{\infty_1, \infty_2, \infty_3\}$ $\{(2i-1,1), (2i-1,2), (2i-1,3)\}$

{(2i-1, 1), (2i, 1), ∞_2}	{(2i-1, 1), (2i, 2), ∞_1}	{(2i, 1), (2i, 2), (2i, 3)}
{(2i-1, 2), (2i, 2), ∞_3}	{(2i-1, 2), (2i, 3), ∞_2}	{(2i-1, 2), (2i, 1), ∞_1}
{(2i -1, 3), (2i, 3), ∞_1}	{(2i-1, 3), (2i, 1), ∞_3}	{(2i-1, 3), (2i, 2), ∞_2}
		{(2i-1, 1), (2i, 3), ∞_3}.

As in Example 1.5.11, for step 2 we can use the commutative quasigroup with holes H of order 10 defined in Example 1.5.6. Again, we will find the triples containing the pairs (i) $(4,1)$ and $(5,2)$ (ii) $(5,1)$ and $(6,2)$ (iii) ∞ and $(8,3)$.

(i) Since $\{4,5\} \notin H$, the answer is the same as Example 1.5.11 (i): $\{(4,1), (10,1), (5,2)\}$.

(ii) Since $\{5,6\} \in H$, we find the triple $\{(2i-1,1), (2i,2), \infty_1\}$ in B_i', so B_3' contains $\{(5,1), (6,2), \infty_1\}$.

(iii) B_i' contains the triple $\{(2i-1,3), (2i,3), \infty_1\}$, so the triple is $\{(7,3), (8,3), \infty_1\}$.

Example 1.5.13 Suppose we construct a $STS(35)$ using the Quasigroup with Holes Construction. Since $35 \equiv 5 \pmod 6$ we use (c) in this construction. In step (1) we need to define B_i'', so let B_i'' contain the following triples (each B_i'' can be any $PBD(11)$ defined on the symbols $\{\infty_1, \ldots, \infty_5\} \cup (\{2i-1, 2i\} \times \{1,2,3\})$ providing that the symbols are named so that $\{\infty_1, \infty_2, \infty_3, \infty_4, \infty_5\}$ is the block of size 5):

{∞_1, ∞_2, ∞_3, ∞_4, ∞_5}	{∞_3, (2i, 2), (2i-1, 3)}
{∞_1, (2i-1, 1), (2i, 1)}	{∞_3, (2i-1, 1), (2i, 3)}
{∞_1, (2i-1, 2), (2i, 2)}	{∞_4, (2i-1, 2), (2i, 3)}
{∞_1, (2i-1, 3), (2i, 3)}	{∞_4, (2i-1, 1), (2i-1, 3)}
{∞_2, (2i-1, 1), (2i, 2)}	{∞_4, (2i, 1), (2i, 2)}
{∞_2, (2i, 1), (2i, 3)}	{∞_5, (2i-1, 1), (2i-1, 2)}
{∞_2, (2i-1, 2), (2i-1, 3)}	{∞_5, (2i, 1), (2i-1, 3)}
{∞_3, (2i, 1), (2i-1, 2)}	{∞_5, (2i, 2), (2i, 3)}.

As in Example 1.5.11, for step 2 we can use the commutative quasigroup with holes H of order 10 defined in Example 1.5.6. Again, we will find the triples containing the pairs (i) $(4,1)$ and $(5,2)$ (ii) $(5,1)$ and $(6,2)$ (iii) ∞ and $(8,3)$.

(i) This answer is the same as Example 1.5.11 (i).

(ii) Since $\{5,6\} \in H$, we find $\{\infty_2, (2i-1,1), (2i,2)\}$ in B_i'', so the triple is $\{\infty_2, (5,1), (6,2)\}$.

(iii) B_i'' contains the triple $\{\infty_1, (2i-1,3), (2i,3)\}$, so the triple is $\{\infty_1, (7,3), (8,3)\}$.

Exercises

1.5.14 Use the Quasigroup with Holes Construction and the quasigroup with holes H of order 6 from Example 1.5.1 to find

(a) a $STS(19)$,
(b) a $STS(21)$, and
(c) a $PBD(23)$ with one block of size 5, the rest of size 3.

1.5.15 A $STS(25)$ is found by using the Quasigroup with Holes Construction, with B_i defined in Example 1.5.11, and the quasigroup with holes H of order 8 found in Example 1.5.1. Find the triple containing the pair of symbols:

(a) (2, 1) and (3, 2)
(b) (3, 1) and (4, 2)
(c) (5, 1) and ∞
(d) (5, 1) and (5, 3)
(e) (4, 1) and (5, 3)
(f) (7, 2) and ∞

1.5.16 A $STS(27)$ is found by using the Quasigroup with Holes Construction, with B_i' defined in Example 1.5.12, and the quasigroup with holes H of order 8 found in Example 1.5.1. Find the triple containing the pair of symbols:

(a) (2, 1) and (4, 3)
(b) (3, 1) and (4, 3)
(c) (6, 2) and (6, 3)
(d) (6, 2) and ∞_3
(e) (6, 2) and (8, 1)
(f) (1, 1) and ∞_2

1.5.17 A $PBD(29)$ is found by using the Quasigroup with Holes Construction, with B_i'' defined in Example 1.5.13, and the quasigroup with holes H of order 8 found in Example 1.5.1. Find the triple containing the pair of symbols:

(a) (2, 1) and ∞_2
(b) (3, 1) and (4, 1)
(c) (3, 3) and (5, 3)
(d) (1, 3) and (2, 1)
(e) (1, 3) and ∞_4
(f) ∞_2 and ∞_4

1.6 The Wilson Construction

Before giving the Wilson Construction for triple systems we will need quite a few preliminaries. A *1-factor* of a graph G is a set of pairwise disjoint edges which partition the vertex set. A *2-factor* of G is a 2-regular spanning subgraph. A *1-factorization* of G is a set of 1-factors which partition the edge set of G.

In what follows we will need a 1-factorization of a very particular graph; the so-called deficiency graph of $(Z_n, +)$ (= the integers modulo n). The *deficiency graph* of $(Z_n, +)$, where $n \equiv 1$ or $5 \pmod 6$, is defined to be the graph $G = (V, E)$ with vertex set $V = Z_n \backslash \{0\}$ and edge set $E = \{\{x, -x\}, \{x, -2x\} \mid x \in Z_n \backslash \{0\}\}$. It is immediately seen that the deficiency graph is 3-regular. What is important for us is the fact that the deficiency graph always has a 1-factorization (into 3 1-factors, of course). We will give a very simple algorithm for doing this, which requires some additional preliminaries.

In this book, a *wheel* is a graph consisting of a cycle of even length and a 1-factor in which the edges join the opposite vertices of the cycle (called the *spokes*). A wheel can be 1-factored into 3 1-factors by taking the alternate edges of the cycle as two of the 1-factors and the spokes for the third.

Example 1.6.1 (Wheel on 8 vertices and 1-factorization).

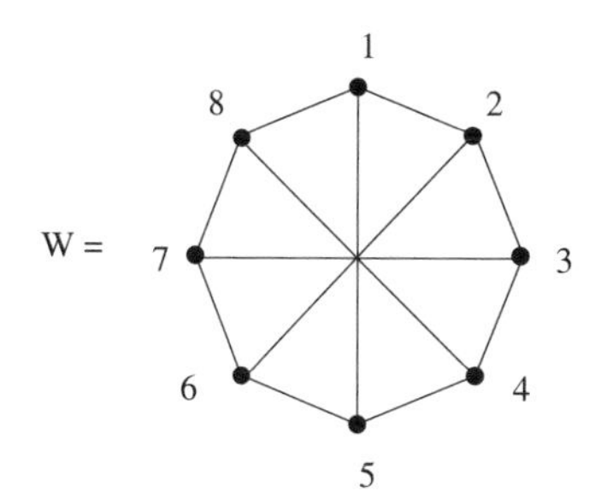

1 2	2 3	1 5
3 4	4 5	2 6
5 6	6 7	3 7
7 8	1 8	4 8

1-factorization of the wheel W

A *biwheel* is a graph consisting of the union of two vertex disjoint cycles C_1 and C_2 of the same length n and a 1-factor F consisting of n edges such that: (i) each edge in F is incident with one vertex in C_1 and one vertex in C_2; and (ii) the mapping $\alpha : V(C_1) \to V(C_2)$ defined by $x\alpha = y$ if and only if $\{x, y\} \in F$ is an *isomorphism* of C_1 onto C_2. Just as was the case for wheels, biwheels can be 1-factored into 3 1-factors. However, the algorithm for doing this is just a bit trickier than the algorithm for wheels (but not by much). To begin with, if C_1 has *even length*, we can take the alternate edges of C_1 and C_2 as two of the 1-factors, and F as the third. If C_1 has *odd length* we proceed in the following manner: (i) partition C_1 into three partial 1-factors P_1, P_2, and P_3; (ii) partition C_2 into 3 partial 1-factors $P_1\alpha$, $P_2\alpha$, and $P_3\alpha$ defined by $\{x\alpha, y\alpha\} \in P_i\alpha$ if and only if $\{x, y\} \in P_i$; and finally (iii) partition F into partial 1-factors F_1, F_2, and F_3 defined by $\{x, x\alpha\} \in F_i$ if and only if $\{a, x\}$ and $\{x, b\}$ in C_1 do *not* belong to P_i. Then each $P_i \cup P_i\alpha \cup F_i$ is a 1-factor and, of course, $\{P_1 \cup P_1\alpha \cup F_1, P_2 \cup P_2\alpha \cup F_2, P_3 \cup P_3\alpha \cup F_3\}$ is a 1-factorization of the biwheel $C_1 \cup C_2 \cup F$.

Example 1.6.2 (Biwheel on 18 vertices and 1-factorization).

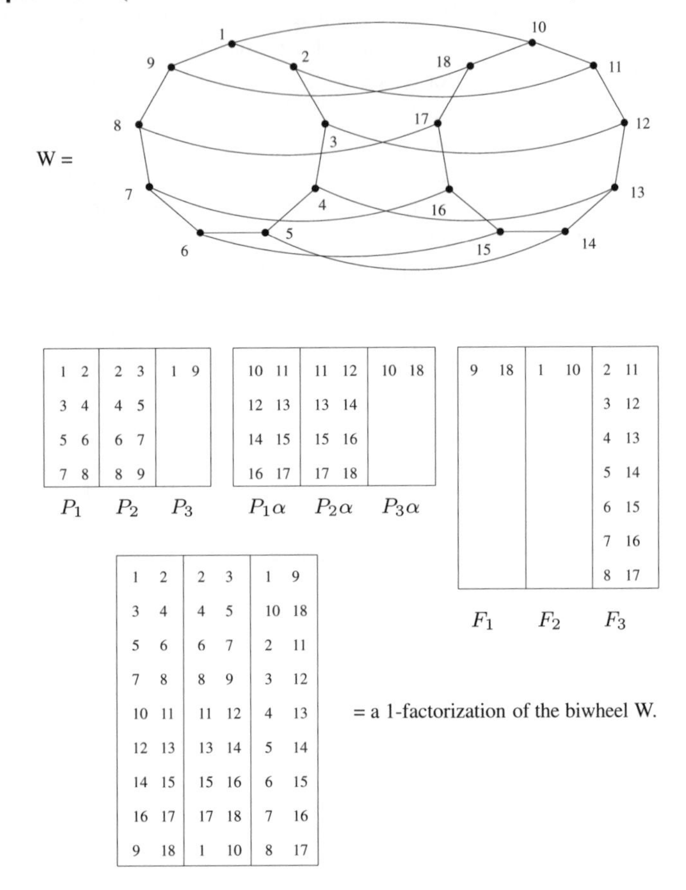

P_1	P_2	P_3
1 2	2 3	1 9
3 4	4 5	
5 6	6 7	
7 8	8 9	

$P_1\alpha$	$P_2\alpha$	$P_3\alpha$
10 11	11 12	10 18
12 13	13 14	
14 15	15 16	
16 17	17 18	

F_1	F_2	F_3
9 18	1 10	2 11
		3 12
		4 13
		5 14
		6 15
		7 16
		8 17

1 2	2 3	1 9
3 4	4 5	10 18
5 6	6 7	2 11
7 8	8 9	3 12
10 11	11 12	4 13
12 13	13 14	5 14
14 15	15 16	6 15
16 17	17 18	7 16
9 18	1 10	8 17

= a 1-factorization of the biwheel W.

The Deficiency Graph Algorithm. Let $m \equiv 1$ or $5 \pmod 6$ and let $G = (V, E)$ be the deficiency graph of $(Z_m, +)$. Let $F = \{\{x, -x\} \mid x \in Z_m \backslash \{0\}\}$ and $T = \{\{x, -2x\} \mid x \in Z_m \backslash \{0\}\}$.

(1) Decompose T into vertex disjoint cycles. (Since T is a 2-factor of G this is always possible.) We consider each such cycle C in turn.

(2) If C is a cycle of *even* length then either every pair of opposite vertices in C are of the form $\{x, -x\}$, or there exists another cycle C^* such that $x \in C$ if and only if $-x \in C^*$ (see Exercise 1.6.9). Regardless, let $F(C)$ be the set of edges of F which cover the vertices of C or of $C \cup C^*$, and form the wheel $C \cup F(C)$ or the biwheel $C \cup C^* \cup F(C)$, as the case may be.

(3) If C is a cycle of *odd* length then there exists a cycle C^* such that $x \in C$ if and only if $-x \in C^*$ (see Exercise 1.6.9). Let $F(C)$ be the set of edges of F which cover the vertices of $C \cup C^*$ and form the biwheel $C \cup C^* \cup F(C)$.

(4) Steps (2) and (3) partition the deficiency graph into vertex disjoint wheels and biwheels. Now, 1-factor each wheel and biwheel and paste these 1-factors together to obtain the 1-factorization of the deficiency graph $G = (V, E)$.

Example 1.6.3 The deficiency graph of $(Z_{55}, +)$ has the following 3 components. For each component we produce a 1-factorization using the Deficiency Graph Algorithm.

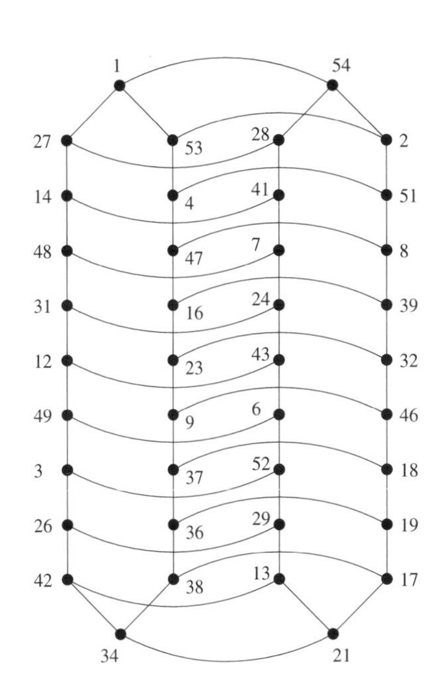

F_0		F_1		F_2	
1	53	53	4	1	54
4	47	47	16	53	2
16	23	23	19	27	28
9	37	37	36	4	51
36	38	38	34	14	41
34	42	42	26	47	8
26	3	3	49	48	7
49	12	12	31	16	39
31	48	48	14	31	24
14	27	27	1	23	32
54	2	2	51	12	43
51	8	8	39	9	46
39	32	32	46	49	6
46	18	18	19	37	18
19	17	17	21	3	52
21	13	13	29	36	19
29	52	52	6	26	29
6	43	43	24	38	17
24	7	7	41	42	13
41	28	28	54	34	21

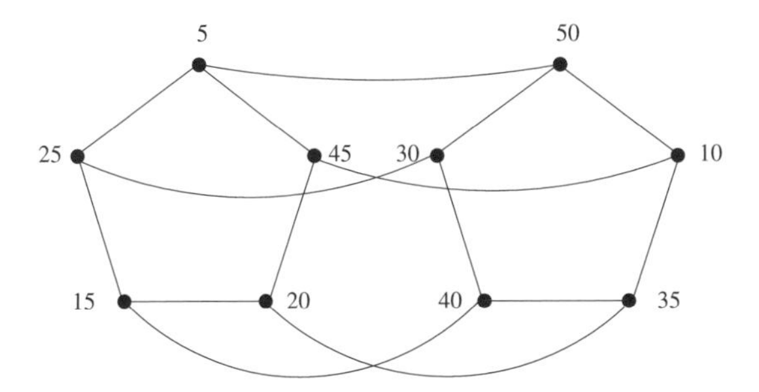

5	45	45	20	25	5
20	15	15	25	30	50
50	10	10	35	45	10
35	40	40	30	20	35
25	30	5	50	15	40

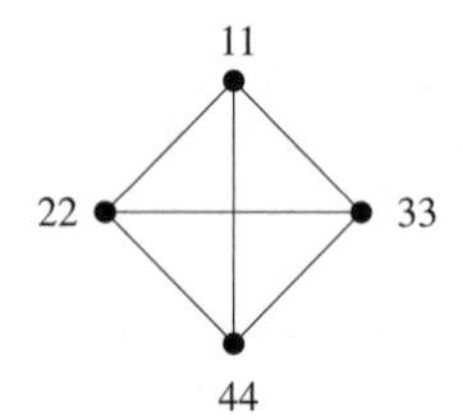

11 33	33 44	11 44
44 22	22 11	22 33

Wilson's Construction ([27] for Steiner triple systems).

Let $v \equiv 1$ or 3 (mod 6) and set $S = \{\infty_1, \infty_2\} \cup Z_{v-2}$. It is *important* to note that $v - 2 \equiv 1$ or 5 (mod 6). Define collections of triples as follows:

Type 1: $T^* = \{\{x, y, z\} \mid x + y + z \equiv 0 \pmod{v-2}\}$, where x, y, z are distinct elements in $Z_{v-2} \backslash \{0\}\}$.

Type 2: The 2-element subsets of $Z_{v-2} \backslash \{0\}$ not covered by a Type 1 triple are precisely the 2-element subsets of the form $\{x, -x\}$ and $\{x, -2x\}$; i.e., the deficiency graph of $(Z_{v-2}, +)$. Let F_0, F_1, and F_2 be a 1-factorization of the deficiency graph and define

$$\begin{cases} T_0 &= \{\{0, x, y\} \mid \{x, y\} \in F_0\}, \\ T_1 &= \{\{\infty_1, x, y\} \mid \{x, y\} \in F_1\}, \text{ and} \\ T_2 &= \{\{\infty_2, x, y\} \mid \{x, y\} \in F_2\}. \end{cases}$$

Type 3: $\{0, \infty_1, \infty_2\}$.

Set $T = T^* \cup T_0 \cup T_1 \cup T_2 \cup \{\{0, \infty_1, \infty_2\}\}$. Then (S, T) is a $STS(v)$.

Example 1.6.4 (Construction of a STS(9) using Wilson's Construction).
$S = \{\infty_1, \infty_2\} \cup Z_7$
Type 1 triples: $\{1, 2, 4\}, \{3, 5, 6\}$
Type 2 triples: the deficiency graph of Z_7 and 1-factorization is given by:

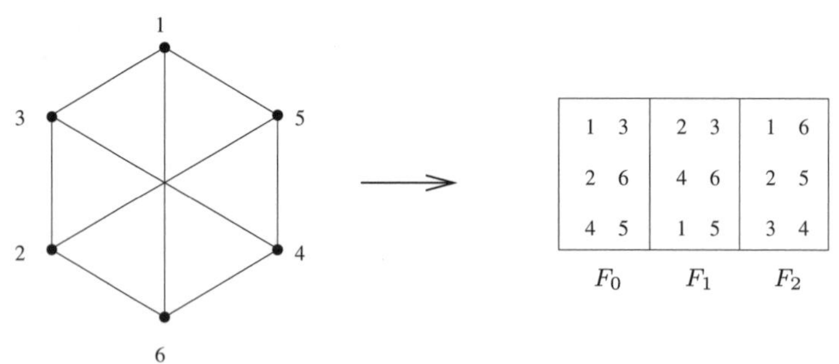

F_0	F_1	F_2
1 3	2 3	1 6
2 6	4 6	2 5
4 5	1 5	3 4

$$\{0, 1, 3\}, \{0, 2, 6\}, \{0, 4, 5\}$$
$$\{\infty_1, 2, 3\}, \{\infty_1, 4, 6\}, \{\infty_1, 1, 5\}$$
$$\{\infty_2, 1, 6\}, \{\infty_2, 2, 5\}, \{\infty_2, 3, 4\}$$

Type 3 triple: $\{0, \infty_1, \infty_2\}$

Then (S, T) is a $STS(9)$ where T is the union of the above triples.

Exercises

1.6.5 Use Wilson's Construction to find a $STS(v)$, where

(a) $v = 7$
(b) $v = 13$.

1.6.6 A $STS(v)$ is to be found using Wilson's Construction. Find the deficiency graph of T^* and a 1-factorization F_0, F_1 and F_2 when

(a) $v = 15$
(b) $v = 19$.

1.6.7 The deficiency graph found in Example 1.6.3 is used to find a $STS(57)$ with F_0, F_1 and F_2 as in the example. Find the triple containing the pair of symbols:

(a) 1 and 27
(b) 5 and 51
(c) 7 and 32
(d) 6 and 7
(e) 2 and ∞_1
(f) 2 and 53
(g) 5 and 45
(h) 11 and 14

1.6.8 Show that the only pairs of symbols in $\{1, 2, \ldots, v-3\}$ that are not in triples in T' are $\{x, -x\}$ and $\{x, -2x\}$ for $1 \leq x \leq v-3$.

1.6.9 Verify the statements in parts (2) and (3) of the Deficiency Graph Algorithm.

1.7 Cyclic Steiner triple systems

An *automorphism* of a STS (S, T) is a bijection $\alpha : S \to S$ such that $t = \{x, y, z\} \in T$ if and only if $t\alpha = \{x\alpha, y\alpha, z\alpha\} \in T$. A $STS(v)$ is *cyclic* if it has an automorphism that is a permutation consisting of a single cycle of length v.

Example 1.7.1 (a) $S = \{1, 2, 3, 4, 5, 6, 7\}, T = \{\{1, 2, 4\}, \{2, 3, 5\}, \{3, 4, 6\}, \{4, 5, 7\}, \{5, 6, 1\}, \{6, 7, 2\}, \{7, 1, 3\}\}$ is a $STS(7)$ which is cyclic, since the permutation $\alpha = (1, 2, 3, 4, 5, 6, 7)$ is an automorphism. (Of course by $(1, 2, 3, 4, 5, 6, 7)$ we mean that $1\alpha = 2, 2\alpha = 3, \ldots, 7\alpha = 1$.) For example, if $t = \{1, 2, 4\}$ then $t\alpha = \{1\alpha, 2\alpha, 4\alpha\} = \{2, 3, 5\}$ which is indeed a triple in T. Similarly, for each $t \in T, t\alpha \in T$.

(b) $S = \{1, 2, 3, 4, 5, 6, 7, 8, 9\}$ and T contains the following triples:

1	2	3	1	4	7	1	5	9	1	6	8
4	5	6	2	5	8	2	6	7	2	4	9
7	8	9	3	6	9	3	4	8	3	5	7

The permutation $\alpha = (1, 2, 3)(4, 5, 6)(7, 8, 9)$ is an automorphism of (S, T). However (S, T) is *not* cyclic since it has no automorphism that is a cycle of length $v = 9$.

It is natural to ask, for which integers v do there exist $STS(v)$ that are cyclic? This question can be answered by solving the following problems that were posed by L. Heffter in 1896 [11]. For each integer v, define a *difference triple* to be a subset of 3 *distinct* elements of $\{1, 2, \ldots, v-1\}$ such that either: (i) their sum is 0 (mod v), or (ii) one element is the sum of the other two (mod v).

In what follows we will always label the elements of a difference triple with x, y and z with the proviso that $x + y = \pm z$ (mod v). Note that for any difference triple defined by (i), we can label the three elements x, y and z in any way we want. However if the difference triple is defined by (ii) then z is uniquely determined.

We remark that x, y, and z are *distinct* and *nonzero*. This is important!

Heffter's Difference Problems

(1) Let $v = 6n + 1$. Is it possible to partition the set $\{1, 2, \ldots, (v-1)/2 = 3n\}$ into difference triples?

(2) Let $v = 6n + 3$. Is it possible to partition the set $\{1, 2, \ldots, (v-1)/2 = 3n + 1\} \setminus \{v/3 = 2n + 1\}$ into difference triples?

Example 1.7.2 A solution to Heffter's first difference problem

(a) when $v = 7$ is $\{\{1, 2, 3\}\}$, and

(b) when $v = 13$ is $\{\{1, 3, 4\}, \{2, 5, 6\}\}$ since $1 + 3 = 4$ and $2 + 5 + 6 = 13 \equiv 0$ (mod v).

A solution to Heffter's second difference problem

(c) when $v = 9$ does not exist, since $\{1, 2, \ldots, (v-1)/2)\} \setminus \{v/3\} = \{1, 2, 4\}$ which is not a difference triple, and

(d) when $v = 15$ is $\{\{1, 3, 4\}, \{2, 6, 7\}\}$.

Exercises

1.7.3 Find a solution to Heffter's difference problem when

(a) $v = 19$
(b) $v = 21$
(c) $v = 25$
(d) $v = 27$
(e) $v = 31$

(f) $v = 33$

If $\{x, y, z\}$ is a difference triple (so $x + y \equiv \pm z \pmod{v}$), we define the corresponding *base block* to be the triple $\{0, x, x + y\}$.

Example 1.7.4 The base blocks corresponding to the difference triples constructed in Example 1.7.2 are

(a) $\{0, 1, 3\}$,

(b) $\{0, 1, 4\}$ and $\{0, 2, 7\}$,

(d) $\{0, 1, 4\}$ and $\{0, 2, 8\}$.

Exercises

1.7.5 For each difference triple constructed in Exercise 1.7.3, find the corresponding base block. (Clearly the base block is determined by the labelling of the difference triple with x, y and z.)

In 1939, Rose Peltesohn [19] solved both of Heffter's Difference Problems, answering each question in the affirmative except for $v = 9$ (for which no solution exists). This solution provides the following theorem, the proof of which shows how to construct a cyclic $STS(v)$ from a solution of Heffter's Difference Problems. (See the appendix for the complete solution.)

Theorem 1.7.6 *For all* $v \equiv 1$ *or* $3 \pmod 6$ $v \neq 9$*, there exists a cyclic* $STS(v)$.

Proof A proof that there is no cyclic $STS(9)$ is left to the reader and so is omitted.

Suppose that $D(v)$ is a set of difference triples that are a solution to Heffter's Difference Problem. Let $B(v)$ be the collection of base blocks obtained from the difference triples in $D(v)$. Define a cyclic $STS(v)$ $(\{0, 1, \ldots, v-1\}, T)$ as follows:

(1) if $v = 6n + 1$ then

$$T = \{\{i, x+i, x+y+i\} | 0 \leq i \leq v-1, \{0, x, x+y\} \in B(v)\}, \text{ and}$$

(2) if $v = 6n + 3$ then

$$T = \{\{i, x+i, x+y+i\} | 0 \leq i \leq v-1, \{0, x, x+y\} \in B(v)\}$$
$$\bigcup \{\{i, 2n+1+i, 4n+2+i\} | 0 \leq i \leq 2n\},$$

where all sums are reduced modulo v. In (2), the set of triples $\{\{i, 2n+1+i, 4n+2+i\} \mid 0 \leq i \leq 2n\}$ is called a *short orbit*, and the triple $\{0, 2n+1, 4n+2\}$ is called *the* base block for the short orbit.

To see that this defines a $STS(v)$, first notice that for any $\{x, y, z\} \in D(v)$, if we consider the difference between each pair of symbols in the corresponding triples $\{i, x+i, x+y+i\}$ in T we get

$$\begin{aligned}(x+i)-i &= x,\\ (x+y+i)-(x+i) &= y, \text{ and}\\ (x+y+i)-i &= x+y \in \{z,-z\},\end{aligned}$$

again doing all arithmetic modulo v.

So for any pair of symbols a and b in $\{0,1,\ldots,v-1\}$, $a-b=d$ and $b-a=-d$, where we can assume that $1 \le d \le (v-1)/2$ (Why?). Then either

(a) $v=6n+3$ and $d=2n+1$, in which case $\{a,b\}$ is in a triple $\{i, 2n+1+i, 4n+2+i\}$ for some i, or

(b) otherwise $d \in \{x,y,z\} \in D(v)$ and $\{a,b\} \in \{i, x+i, x+y+i\}$ for some i.

□

Example 1.7.7 In Example 1.7.2, the solution

$$D(13) = \{\{1,3,4\},\{2,5,6\}\}$$

to Heffter's Difference Problems was found. Corresponding to $\{1,3,4\}$ we get the base block $\{0,1,4\}$ by choosing $x=1, y=3$ and $z=4$ (so $x+y=z$), and corresponding to $\{2,5,6\}$ we get the base block $\{0,2,7\}$ by choosing $x=2, y=5$ and $z=6$ (so $x+y=-z$). Then

$$T = \{\{i,1+i,4+i\},\{i,2+i,7+i\} | 0 \le i \le 12\}$$

defines the triples in a cyclic $STS(13)$.

Notice that we could choose x, y and z differently in each case. We only have to make sure that $x+y=\pm z$. So for example, for $\{1,3,4\}$ choose $x=3, y=1$ and $z=4$, and for $\{2,5,6\}$ choose $x=6, y=5$ and $z=2$, so that then

$$T = \{\{i,3+i,4+i\},\{i,6+i,11+i\}\}$$

gives a different cyclic $STS(13)$.

Again, in Example 1.7.2, the solution

$$D(15) = \{\{1,3,4\},\{2,6,7\}\}$$

was obtained. The difference triples in $D(15)$ can yield base blocks $\{0,1,4\}$ and $\{0,2,8\}$ (by appropriately choosing x, y and z), so

$$\begin{aligned}T &= \{\{i,1+i,4+i\},\{i,2+i,8+i\} | 0 \le i \le 14\} \bigcup\\ &\quad \{\{i,5+i,10+i\} | 0 \le i \le 4\}\end{aligned}$$

define the triples in a cyclic $STS(15)$. Don't forget the blocks in the short orbit!

Exercises

1.7.8 Use the solutions to Heffter's Difference Problems found in Exercise 1.7.3 to form base blocks for a cyclic $STS(v)$.

1.7.9 The base blocks $\{0, 3, 4\}$, $\{0, 6, 8\}$ and $\{0, 5, 10\}$ can be used to form a cyclic $STS(15)$ (S, T). Find the triples in T containing the symbols

(a) 7 and 14

(b) 0 and 2

(c) 7 and 12

(d) 1 and 12

(e) 0 and 7

(f) 6 and 10.

Chapter 2

λ-Fold Triple Systems

2.1 Triple systems of index $\lambda > 1$.

A *λ-fold triple system* (or a triple system of *index* λ) is a pair (S, T), where S is a finite set and T is a collection of 3-element subsets of S called *triples* such that each pair of distinct elements of S belongs to *exactly* λ triples of T. So, a Steiner triple system is a 1-fold triple system (or a triple system of index $\lambda = 1$). As with Steiner triple systems, the *order* of a λ-fold triple system (S, T) is the number $|S|$. Just as with Steiner triple systems we can think of a λ-fold triple system as a decomposition of λK_v, the graph with v vertices in which each vertex is joined by λ edges, into edge disjoint triangles.

It is *important* to note that nothing in the definition of a λ-fold triple system requires that the triples be *distinct*. That is, in a given example, a triple may be repeated as many as λ times.

Figure 2.1 is a pictorial representation of a 2-fold triple system.

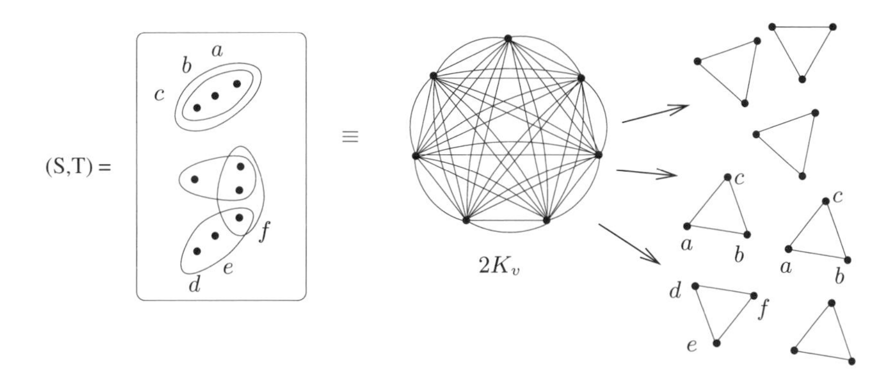

Figure 2.1: *2-fold triple system.*

Example 2.1.1 The following is an example of a 3-fold triple system of order 5. Notice that there are *no repeated* triples.

$$S = \{1, 2, 3, 4, 5\}$$

and T consists of the following triples:

1 2 3	1 2 4
2 3 4	2 3 5
3 4 5	3 4 1
4 5 1	4 5 2
5 1 2	5 1 3

Example 2.1.2 The following is an example of a 4-fold triple system of order 4. Notice that each triple is *repeated* twice.

$$S = \{1, 2, 3, 4\},$$

and T consists of the following triples:

1 2 3	1 2 3
1 2 4	1 2 4
1 3 4	1 3 4
2 3 4	2 3 4

Example 2.1.3 The following is an example of a 2-fold triple system of order 7. Notice that the triples containing 7 are repeated twice, but the remaining triples are distinct; i.e., occur exactly once.

$$S = \{1, 2, 3, 4, 5, 6, 7\},$$

and T consists of the following triples;

1 2 7	1 2 7
3 4 7	3 4 7
5 6 7	5 6 7
1 3 5	2 4 6
1 4 6	2 3 5
2 3 6	1 3 6
2 4 5	1 4 5

The above examples illustrate the fact that λ-fold triple systems may or may not have repeated triples. However, in what follows we will not be concerned with whether or not λ-fold triple systems have repeated triples. We will be concerned with the existence problem only.

We will solve the existence problem for $\lambda = 1, 2, 3$, and 6 *only*. All other values of λ can be obtained by "pasting" together smaller values of λ (see Exercise 2.5.7).

Quite naturally, the place to begin is with $\lambda = 2$. However, in order to construct 2-fold triple systems we must first construct idempotent quasigroups of *every order* $n \neq 2$. We already know how to construct *idempotent commutative* quasigroups of every *odd* order. So half of the problem is solved. However the construction of idempotent quasigroups of even order is not completely obvious.

2.2 The existence of idempotent latin squares

A *transversal* T of a latin square of order n on the symbols $\{1, 2, \ldots, n\}$ is a set of n cells, exactly one cell from each row and each column, such that each of the symbols in $\{1, 2, 3, \ldots, n\}$ occurs in a cell of T.

Example 2.2.1 Let

$L =$

1	3	4	2
4	2	1	3
2	4	3	1
3	1	2	4

Then L has many transversals including

$$\begin{aligned} T_1 &= \{(1,1), (2,2), (3,3), (4,4)\}, \\ T_2 &= \{(1,3), (2,4), (3,1), (4,2)\}, \text{ and} \\ T_3 &= \{(1,1), (2,4), (3,2), (4,3)\}. \end{aligned}$$

Stripping a transversal. If $(\{1, 2, \ldots, n-1\}, \circ)$ is a quasigroup of order $n-1$ that contains a transversal T, then this quasigroup can be used to produce a quasigroup $(\{1, 2, \ldots, n\}, *)$ by *stripping the transversal* T as follows.

(1) For each $(i, j) \in T$, define

$$i * n = n * j = i \circ j, \text{ and } i * j = n.$$

(2) For each $(i, j) \notin T, 1 \leq i, j \leq n-1$ define

$$i * j = i \circ j.$$

(3) Define $n * n = n$.

Example 2.2.2 Using the quasigroup

∘	1	2	3	4
1	1	3	4	2
2	4	2	1	3
3	2	4	3	1
4	3	1	2	4

defined in Example 2.2.1, by stripping T_2 (the cells in T_2 are shaded) we get the quasigroup

*	1	2	3	4	5
1	1	3	5	2	4
2	4	2	1	5	3
3	5	4	3	1	2
4	3	5	2	4	1
5	2	1	4	3	5

and stripping the transversal T_3 (defined in Example 2.2.1 gives the quasigroup

*	1	2	3	4	5
1	5	3	4	2	1
2	4	2	1	5	3
3	2	5	3	1	4
4	3	1	5	4	2
5	1	4	2	3	5

(You can think of the symbols in the cells of the transversal migrating across to the last column and down to the last row, and being replaced in the vacated cells with the new symbol n.)

Theorem 2.2.3 *For all $n \neq 2$ there exists an idempotent quasigroup of order n.*

Proof If n is odd we can rename the addition table for the integers modulo n to form an idempotent latin square of order n (see Exercise 1.2.3). Now $T = \{(1,2),(2,3),\ldots,(n-1,n),(n,1)\}$ is a transversal of the addition table for the integers modulo n when n is odd (see Exercise 2.2.5), and so is also a transversal of the idempotent quasigroup obtained by renaming it. Therefore we can form an idempotent quasigroup of order $n+1$ by stripping the transversal T. □

Example 2.2.4 For $n = 5$ we form the idempotent quasigroup defined in the proof of Theorem 2.2.3 as follows.

∘	1	2	3	4	5
1	1	4	2	5	3
2	4	2	5	3	1
3	2	5	3	1	4
4	5	3	1	4	2
5	3	1	4	2	5

Then stripping the transversal $T = \{(1,2),(2,3),(3,4),(4,5),(5,1)\}$ gives the following idempotent quasigroup of order 6.

*	1	2	3	4	5	6
1	1	6	2	5	3	4
2	4	2	6	3	1	5
3	2	5	3	6	4	1
4	5	3	1	4	6	2
5	6	1	4	2	5	3
6	3	4	5	1	2	6

Exercises

2.2.5 Let n be odd. Prove that $\{(1,2),(2,3),\ldots,(n-1,n),(n,1)\}$ is a transversal of the addition table for the integers modulo n.

2.2.6 Construct idempotent quasigroups of orders $6, 7, 8, 9$, and 10.

2.3 2-Fold triple systems

2.3.1 Constructing 2-fold triple systems

Example 2.3.1 The following provides 2-fold triple systems of orders $3, 4, 7, 9$ and 10.

(a) $S = \{1,2,3\}, T = \{\{1,2,3\},\{1,2,3\}\}$ (everybody's favorite).

(b) $S = \{1,2,3,4\}, T = \{\{1,2,4\},\{1,2,3\},\{1,3,4\},\{2,3,4\}\}$.

(c) $S = \{1,2,3,4,5,6,7\}$ and T contains the following triples:

1	2	4	1	2	6
2	3	5	2	3	7
3	4	6	3	4	1
4	5	7	4	5	2
5	6	1	5	6	3
6	7	2	6	7	4
7	1	3	7	1	5

(d) $S = \{1,2,3,4,5,6,7\}$ and T contains the following triples:

1	2	4	1	2	4
2	3	5	2	3	5
3	4	6	3	4	6
4	5	7	4	5	7
5	6	1	5	6	1
6	7	2	6	7	2
7	1	3	7	1	3

(e) Any triple system of order 9 with each triple repeated twice.

(f) $S = \{1,2,3,4,5,6,7,8,9,10\}$ and T contains the following triples:

1	2	3	2	4	6	3	6	10
1	2	8	2	4	10	3	7	10
1	3	9	2	5	7	4	5	8
1	4	5	2	5	9	4	6	9
1	4	10	2	6	10	4	7	9
1	5	10	2	7	9	5	6	8
1	6	7	3	4	7	5	7	10
1	6	9	3	4	8	6	7	8
1	7	8	3	5	6	8	9	10
2	3	8	3	5	9	8	9	10

Just as with Steiner triple systems (that is, triple systems of index $\lambda = 1$) we can think of a 2-fold triple system as a decomposition of $2K_v$, the graph with v vertices in which each pair of vertices is joined by 2 edges, into edge disjoint triangles.

Exercises

2.3.2 If (S, T) is a 2-fold triple system of order v, then show that $|T| = v(v-1)/3$.

2.3.3 Deduce that a *necessary* condition for the existence of a 2-fold triple system of order v is $v \equiv 0$ or 1 (mod 3).

2.3.4 Use trial and error to construct a 2-fold triple system of order 6.

2.3.5 Let S be a set of size v and T a collection of 3-element subsets of S such that

(a) every pair of elements of S belongs to *at least* two triples of T, and

(b) $|T| \leq v(v-1)/3$.

Show that (S, T) is a 2-fold triple system of order v. (Compare this to Exercise 1.1.4.)

Remark 2.3.6 Exercise 2.3.5 gives an easy technique for showing that (S, T) is a 2-fold triple system. Simply count the number of triples; and if this number is less than or equal to the correct number of triples, and every pair in in at least two triples, then every pair is in *exactly* two triples!

The *spectrum* for 2-fold triple systems is defined to be the set of integers v for which there exists a 2-fold triple system of order v.

Theorem 2.3.7 *The spectrum for 2-fold triple systems is* precisely *the set of all $v \equiv 0$ or 1 (mod 3).*

The following two constructions provide a proof of the above theorem.

The $3n$ Construction. Let $(Q, \circ)$ be an idempotent (not necessarily commutative) quasigroup of order n and set $S = Q \times \{1, 2, 3\}$. Define a collection of triples T as follows:

Type 1: $\{(x, 1), (x, 2), (x, 3)\}$ occurs exactly twice in T for every $x \in Q$; and

Type 2: if $x \neq y$, the six triples $\{(x, 1), (y, 1), (x \circ y, 2)\}$, $\{(y, 1), (x, 1), (y \circ x, 2)\}$, $\{(x, 2), (y, 2), (x \circ y, 3)\}$, $\{(y, 2), (x, 2), (y \circ x, 3)\}$, $\{(x, 3), (y, 3), (x \circ y, 1)\}$, $\{(y, 3), (x, 3), (y \circ x, 1)\}$ belong to T.

Then (S, T) is a 2-fold triple system of order $3n$.

Figure 2.2 represents this construction graphically.

Type 1 triples : repeat each of these vertical triples.

Type 2 triples.

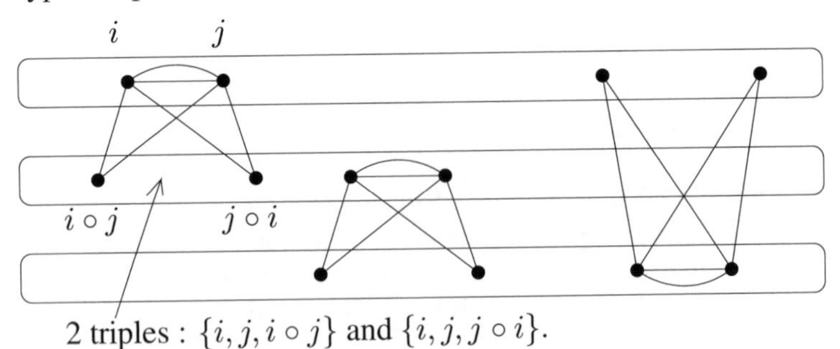

Figure 2.2: *The $3n$ Construction of 2-fold triple systems.*

The $3n+1$ Construction. Let $(Q, \circ)$ be an idempotent (not necessarily commutative) quasigroup of order n and set $S = \{\infty\} \cup (Q \times \{1, 2, 3\})$. Define a collection of triples T as follows:

Type 1: The four triples $\{\infty, (x,1), (x,2)\}$, $\{\infty, (x,2), (x,3)\}$, $\{\infty, (x,1), (x,3)\}$, $\{(x,1), (x,2), (x,3)\}$ belong to T for every $x \in Q$ (note: these 4 triples form a 2-fold triple system of order 4); and

Type 2: if $x \neq y$, the six triples $\{(x,1), (y,1), (x \circ y, 2)\}$, $\{(y,1), (x,1), (y \circ x, 2)\}$, $\{(x,2), (y,2), (x\circ, y, 3)\}$, $\{(y,2), (x,2), (y \circ x, 3)\}$, $\{(x,3), (y,3), (x \circ y, 1)\}$, $\{(y,3), (x,3), (y \circ x, 1)\}$ belong to T.

Then (S, T) is a 2-fold triple system of order $3n+1$.

Type 1 triples :

For each $i \in Q$ define a 2-fold triple system on $\{\infty, (i,1), (i,2), (i,3)\}$.

Type 2 triples.

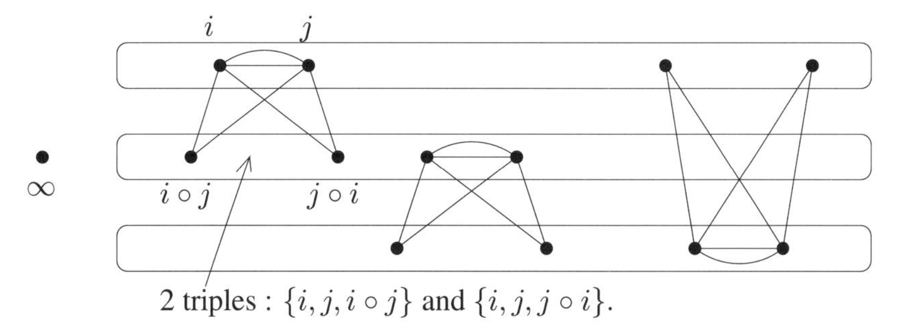

Figure 2.3: *The $3n+1$ Construction of 2-fold triple systems.*

Example 2.3.8 We use the following idempotent quasigroup of order 4 to construct 2-fold triple systems of orders 12 and 13 by using the $3n$ and $3n+1$ Constructions.

$\circ$	1	2	3	4
1	1	3	4	2
2	4	2	1	3
3	2	4	3	1
4	3	1	2	4

Order 12.

Type 1: $\{(1,1),(1,2),(1,3)\}$, $\{(1,1),(1,2),(1,3)\}$, $\{(2,1),(2,2),(2,3)\}$, $\{(2,1),(2,2),(2,3)\}$, $\{(3,1),(3,2),(3,3)\}$, $\{(3,1),(3,2),(3,3)\}$, $\{(4,1),(4,2),(4,3)\}$, $\{(4,1),(4,2),(4,3)\}$.

Type 2: $\{(1,1),(2,1),(3,2)\}$, $\{(2,1),(1,1),(4,2)\}$, $\{(1,1),(3,1),(4,2)\}$, $\{(3,1),(1,1),(2,2)\}$, $\{(1,1),(4,1),(2,2)\}$, $\{(4,1),(1,1),(3,2)\}$, $\{(2,1),(3,1),(1,2)\}$, $\{(3,1),(2,1),(4,2)\}$, $\{(2,1),(4,1),(3,2)\}$, $\{(4,1),(2,1),(1,2)\}$, $\{(3,1),(4,1),(1,2)\}$, $\{(4,1),(3,1),(2,2)\}$; plus two more copies of these triples with the second coordinates increased by 1 and then by 2 (mod 3).

Order 13.

Type 1: $\{\infty,(1,1),(1,3)\}$, $\{\infty,(1,2),(1,1)\}$, $\{\infty,(1,3),(1,2)\}$, $\{(1,1),(1,2),(1,3)\}$;
$\{\infty,(2,1),(2,3)\}$, $\{\infty,(2,2),(2,1)\}$, $\{\infty,(2,3),(2,2)\}$, $\{(2,1),(2,2),(2,3)\}$;
$\{\infty,(3,1),(3,3)\}$, $\{\infty,(3,2),(3,1)\}$, $\{\infty,(3,3),(3,2)\}$, $\{(3,1),(3,2),(3,3)\}$;
$\{\infty,(4,1),(4,3)\}$, $\{\infty,(4,2),(4,1)\}$, $\{\infty,(4,3),(4,2)\}$, $\{(4,1),(4,2),(4,3)\}$.

Type 2: The same as the type 2 triples for the example of order 12 above.

Exercises

2.3.9 Let (S, T) be the 2-fold triple system of order 22 constructed from the accompanying idempotent quasigroup of order 7.

$\circ$	1	2	3	4	5	6	7
1	1	6	4	2	7	5	3
2	4	2	7	5	3	1	6
3	7	5	3	1	6	4	2
4	3	1	6	4	2	7	5
5	6	4	2	7	5	3	1
6	2	7	5	3	1	6	4
7	5	3	1	6	4	2	7

Find the *two triples* containing each of the following pairs.

(a) ∞ and $(7, 2)$
(b) $(2, 2)$ and $(6, 2)$
(c) $(3, 3)$ and $(4, 1)$
(d) $(2, 4)$ and $(3, 4)$
(e) $(6, 2)$ and $(5, 1)$

2.4 $\lambda = 3$ and 6

When $\lambda = 3$ or 6, λ-fold triple systems are particularly easy to construct. In fact we have already constructed such designs!

Example 2.4.1 (a) (3-fold triple system of order 7)
$S = \{1, 2, 3, 4, 5, 6, 7\}$, and T consists of the following triples:

1 2 5	2 4 3	3 7 5
1 3 2	2 5 7	4 5 1
1 4 6	2 6 4	4 6 5
1 5 3	2 7 1	4 7 2
1 6 7	3 4 7	5 6 2
1 7 4	3 5 4	5 7 6
2 3 6	3 6 1	6 7 3

(b) (6-fold triple system of order 4)
$S = \{1, 2, 3, 4\}$, and T consists of the following triples:

1 2 3	2 1 4	3 1 2	4 1 3
1 3 4	2 3 1	3 2 4	4 2 1
1 4 2	2 4 3	3 4 1	4 3 2

Exercises

2.4.2 If (S, T) is a λ-fold triple system, then $|T| = \lambda v(v-1)/6$.

2.4.3 A *necessary* condition for the existence of a 3-fold triple system of order v is that v is odd.

2.4.4 A *necessary* condition for the existence of a 6-fold triple system of order v is that v is a positive integer $\neq 2$.

2.4.5 Let S be a set of size v and T a collection of 3-element subsets of S such that

(a) every pair of elements of S belongs to at least λ triples of T, and
(b) $|T| \leq \lambda v(v-1)/6$

Prove that (S, T) is a λ-fold triple system of order v.

Remark 2.4.6 Exercise 2.4.5 gives an easy technique for showing that (S, T) is a λ-fold triple system. First, count the number of triples; and if this number is less than or equal to the correct number, and every pair is in at least λ triples, then every pair is in *exactly* λ triples.

Theorem 2.4.7 *The spectrum for 3-fold triple systems is* precisely *the set of all odd integers $v \geq 1$. The spectrum for 6-fold triple systems is* precisely *the set of all positive integers $v \neq 2$.*

Proof The following two constructions (really observations) provide a proof of this theorem.

The 3-fold Construction. Let $(Q, \circ)$ be an idempotent commutative quasigroup of order v. Let $T = \{\{a, b, a \circ b\} \mid a < b \in Q\}$. Then (Q, T) is a 3-fold triple system of order v.

The 6-fold Construction. Let $(Q, \circ)$ be an idempotent quasigroup of order v. Let $T = \{\{a, b, a \circ b\} \mid a \neq b \in Q\}$. Then (Q, T) is a 6-fold triple system of order v.

The fact that the above two constructions do in fact produce 3-fold and 6-fold triple systems respectively is left as a very easy exercise (see Exercise 2.4.9).

□

Example 2.4.8 The accompanying quasigroups give the 3-fold triple system of order 7 and the 6-fold triple systems of order 4 in Example 2.4.1

$\circ_1$	1	2	3	4	5	6	7
1	1	5	2	6	3	7	4
2	5	2	6	3	7	4	1
3	2	6	3	7	4	1	5
4	6	3	7	4	1	5	2
5	3	7	4	1	5	2	6
6	7	4	1	5	2	6	3
7	4	1	5	2	6	3	7

$\circ_2$	1	2	3	4
1	1	3	4	2
2	4	2	1	3
3	2	4	3	1
4	3	1	2	4

Exercises

2.4.9 Prove that the constructions given in Theorem 2.4.7 do in fact produce 3-fold and 6-fold triple systems.

2.4.10 Let $(\{1,\ldots,5\},T)$ be the 3-fold triple system constructed from

1	3	5	2	4
3	2	4	5	1
5	4	3	1	2
2	5	1	4	3
4	1	2	3	5

Using the 3-fold Construction. Find all the triples in T containing the pair of symbols:

(a) 1 and 3
(b) 2 and 5
(c) 4 and 5

2.4.11 Let $(\{1,\ldots,5\},T)$ be the 6-fold triple system constructed from

1	3	4	5	2
4	2	5	3	1
5	1	3	2	4
2	5	1	4	3
3	4	2	1	5

using the 6-fold Construction. Find all the triples in T containing the pair of symbols:

(a) 1 and 5
(b) 2 and 4
(c) 3 and 5

2.5 λ-Fold triple systems in general.

In Sections 2.3 and 2.4 we have determined the spectrum for λ-fold triple systems for $\lambda = 2, 3$, and 6. If we don't mind repeated triples (and we don't!) these results are sufficient, along with the spectrum for Steiner triple systems ($\lambda = 1$), to determine the spectrum for λ-fold triple systems for any λ. The accompanying table gives a summary of the spectrum problem for λ-fold triple systems in general.

λ	spectrum of λ-fold triple systems
0 (mod 6)	all $v \neq 2$
1 or 5 (mod 6)	all $v \equiv 1$ or 3 (mod 6)
2 or 4 (mod 6)	all $v \equiv 0$ or 1 (mod 3)
3 (mod 6)	all odd v

Theorem 2.5.1 *The above table provides necessary and sufficient conditions for the existence of a λ-fold triple system of order v.* □

The verification of Theorem 2.5.1 is left as an exercise.

Example 2.5.2 (8-fold triple system of order 4).

$$\begin{aligned} S &= \{1,2,3,4\} \\ T_1 &= \left\{ \begin{array}{cccc} 123 & 214 & 312 & 413 \\ 134 & 231 & 324 & 421 \\ 142 & 243 & 341 & 432 \end{array} \right. \\ T_2 &= \{123,\ 124,\ 134,\ 234\} \end{aligned}$$

Note that (S, T_1) is a 6-fold triple system of order 4 *and* (S, T_2) is a 2-fold triple system of order 4 as a consequence $(S, T_1 \cup T_2)$ is an 8-fold triple system of order 4.

Exercises

2.5.3 If $\lambda \equiv 1$ or 5 (mod 6), and if there exists a λ-fold triple system of order v, prove that $v \equiv 1$ or 3 (mod 6). (Hint: $|T| = \lambda v(v-1)/6$.)

2.5.4 If $\lambda \equiv 2$ or 4 (mod 6), and if there exists a λ-fold triple system of order v, prove that $v \equiv 0$ or 1 (mod 3).

2.5.5 If $\lambda \equiv 3$ (mod 6), and if there exists a λ-fold triple system of order v, prove that v is odd.

2.5.6 Construct the following λ-fold triple systems:

(a) 9-fold triple system of order 5.
(b) 10-fold triple system of order 6.
(c) 5-fold triple of order 9.
(d) 12-fold triple system of order 5.

2.5.7 The necessary conditions in the above table are sufficient for the existence of a λ-fold triple system of order v. (Hint: paste together copies of λ-fold triple systems of smaller index.)

Chapter 3

Maximum Packings and Minimum Coverings

3.1 The general problem

As we are well aware by now, it is *impossible* to construct a Steiner triple system of order v when $v \not\equiv 1$ or 3 (mod 6). A quite natural question to ask then, is just how "close" can we come to constructing a triple system for $v \equiv 0, 2, 4$, or 5 (mod 6). Obviously, in order to attack this problem we must first define what we mean by "close". When $v \equiv 5$ (mod 6), in Section 1.4 we defined "close" to mean one block of size 5 and the remaining blocks of size 3. In what follows we will define "close" in two different ways. So, here goes!

The following two definitions are best described using graph theoretic vernacular.

A *packing* of the complete graph K_v with triangles is a triple (S, T, L), where S is the vertex set of K_v, T is a collection of edge disjoint triangles from the edge set of K_v, and L is the collection of edges in K_v *not* belonging to one of the triangles of T. The collection of edges L is called the *leave*. If $|T|$ is as *large* as possible, or equivalently if $|L|$ is as *small* as possible, then (S, T, L) is called a *maximum packing with triangles* (MPT), or simply a *maximum packing* of order v. So, for example, a Steiner triple system is a maximum packing with leave the *empty set*. Lots of other examples are possible. Here are a few!

Example 3.1.1 (Maximum packings)

(a) (S_1, T_1, L_1) is a maximum packing of order 6.

$$\begin{aligned} S_1 &= \{1, 2, 3, 4, 5, 6\} \\ T_1 &= \{\{1, 2, 4\}, \{2, 3, 5\}, \{3, 4, 6\}, \{1, 5, 6\}\}, \\ L_1 &= \{\{1, 3\}, \{2, 6\}, \{4, 5\}\}. \end{aligned}$$

(b) (S_2, T_2, L_2) is a maximum packing of order 10.

$$\begin{aligned} S_2 &= \{1,2,3,4,5,6,7,8,9,10\}, \\ T_2 &= \{\{2,3,4\},\{1,6,7\},\{1,8,9\},\{1,5,10\},\{2,6,9\},\{2,5,7\}, \\ &\quad \{2,8,10\},\{3,5,6\},\{3,7,8\},\{3,9,10\},\{4,6,10\},\{4,7,9\}, \\ &\quad \{4,5,8\}\}, \\ L_2 &= \{\{1,2\},\{1,3\},\{1,4\},\{6,8\},\{7,10\},\{5,9\}\}. \end{aligned}$$

(c) (S_3, T_3, L_3) is a maximum packing of order 11.

$$\begin{aligned} S_3 &= \{1,2,3,4,5,6,7,8,9,10,11\}, \\ T_3 &= \{\{1,3,5\},\{2,4,5\},\{1,6,7\},\{1,8,9\},\{1,10,11\}, \\ &\quad \{2,6,9\},\{2,7,11\},\{2,8,10\},\{3,6,11\},\{3,7,8\},\{3,9,10\}, \\ &\quad \{4,6,10\},\{4,7,9\},\{4,8,11\},\{5,6,8\},\{5,7,10\},\{5,9,11\}\}, \\ L_3 &= \{\{1,2\},\{2,3\},\{3,4\},\{4,1\}\}. \end{aligned}$$

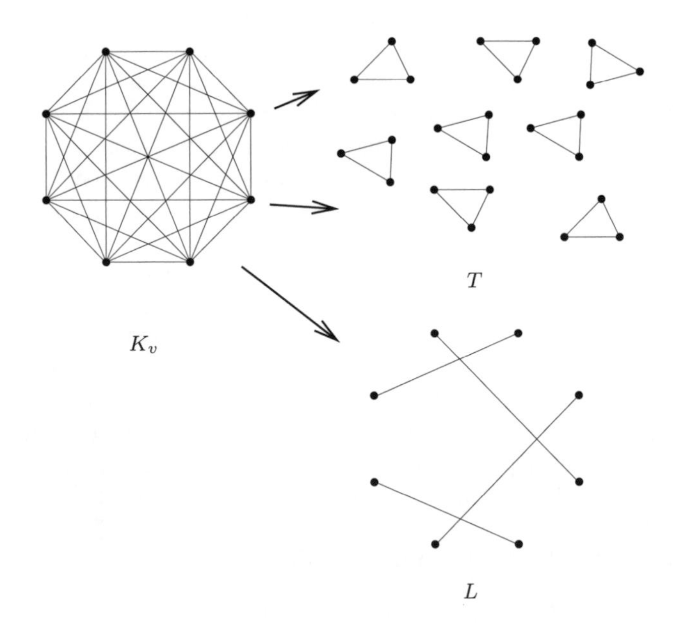

Figure 3.1: *A maximum packing with leave being a* 1*-factor.*

It turns out that maximum packings of the same order all have one thing in common; the leave! In particular if (S, T, L) is a MPT of order v, then the leave is

(i) a 1-factor if $v \equiv 0$ or $2 \pmod 6$,

(ii) a 4-cycle if $v \equiv 5 \pmod 6$,

(iii) a *tripole*, that is a spanning graph with each vertex having odd degree and containing $(v+2)/2$ edges, if $v \equiv 4 \pmod 6$, and of course,

(iv) the empty set if $v \equiv 1$ or $3 \pmod 6$.

We will prove (i), (ii), (iii), and (iv) in Section 3.2.

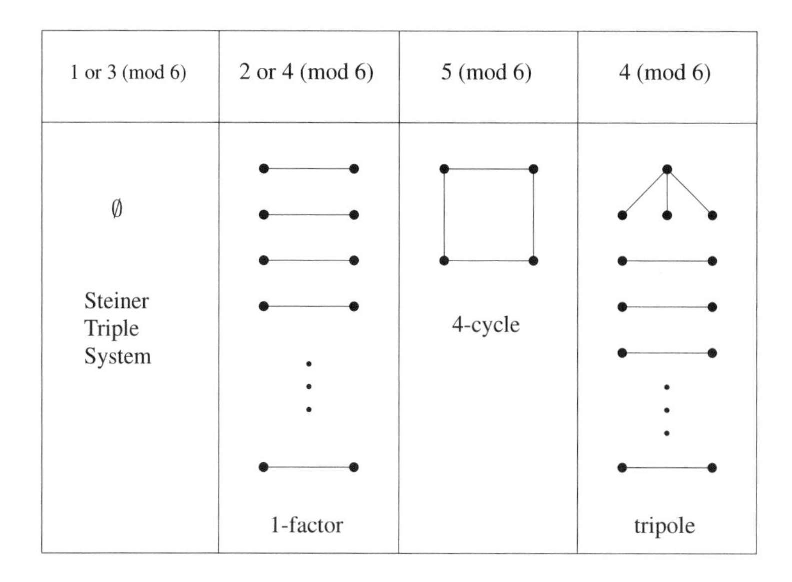

Figure 3.2: *Leaves of maximum packings.*

The examples in Examples 3.1.1 illustrate the three possibilities for $v \equiv 0, 2, 4,$ or $5 \pmod 6$: the leave in (a) is a 1-factor, in (b) is a tripole, and in (c) is a 4-cycle.

We now present our second definition of being "close" to a Steiner triple system.

A *covering* of the complete graph K_v with triangles is a triple (S, T, P), where S is the vertex set of K_v, P is a subset of the edge set of λK_v based on S (λK_v is the graph in which each pair of vertices is joined by λ edges), and T is a collection of edge disjoint triangles which partitions the union of P and the edge set of K_v. The collection of edges P is called the *padding* and the number v the *order* of the covering (S, T, P). So that there is no confusion, we emphasize that an edge $\{a, b\}$ belongs to exactly $x + 1$ triangles of T, where x is the number of times $\{a, b\}$ belongs to P. If $|P|$ is as small as possible the covering (S, T, P) is called a *minimum covering with triangles* (MCT), or more simply a *minimum covering* of order v. So a Steiner triple system is a minimum covering with padding $P = \emptyset$.

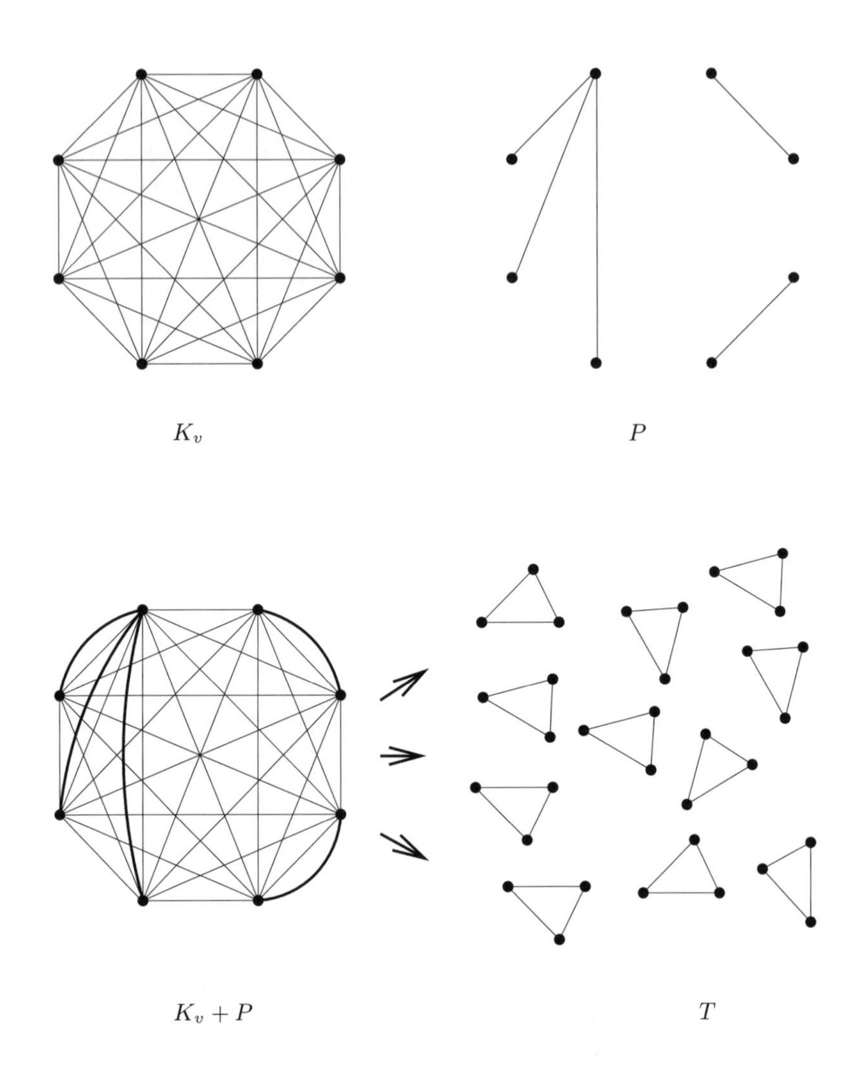

Figure 3.3: *A minimum covering with padding a tripole.*

Example 3.1.2 (Minimum coverings)

(a) (S_1, T_1, P_1) is a minimum covering of order 5.

$$\begin{aligned} S_1 &= \{1,2,3,4,5\}, \\ T_1 &= \{\{1,2,4\},\{1,2,3\},\{1,2,5\},\{3,4,5\}\}, \\ P_1 &= \{\{1,2\},\{1,2\}\}. \end{aligned}$$

(b) (S_2, T_2, P_2) is a minimum covering of order 6.

$$S_2 = \{1,2,3,4,5,6\},$$
$$T_2 = \{\{1,2,3\},\{1,2,4\},\{3,4,5\},\{3,4,6\},\{2,5,6\},\{1,5,6\}\},$$
$$P_2 = \{\{1,2\},\{3,4\},\{5,6\}\}$$

(c) (S_3, T_3, P_3) is a minimum covering of order 8.

$$S_3 = \{1,2,3,4,5,6,7,8\},$$
$$T_3 = \{\{1,2,7\},\{1,4,5\},\{3,5,6\},\{1,2,3\},\{2,4,8\},\{5,7,8\}, \{1,3,8\},\{2,5,6\},\{6,7,8\},\{1,4,6\},\{3,4,7\}\},$$
$$P_3 = \{\{1,2\},\{1,3\},\{1,4\},\{5,6\},\{7,8\}\}.$$

Just as was the case for maximum packings, the padding of a minimum covering is determined by its order. In particular, if (S, T, P) is a MCT of order v, then the padding is

(i) a 1-factor if $v \equiv 0 \pmod 6$,

(ii) a tripole if $v \equiv 2$ or $4 \pmod 6$,

(iii) a double edge $= \{\{a,b\},\{a,b\}\}$ if $v \equiv 5 \pmod 6$, and of course,

(iv) the empty set if $v \equiv 1$ or $3 \pmod 6$.

As with maximum packings we will prove (i), (ii), (iii), and (iv) in Section 3.3.

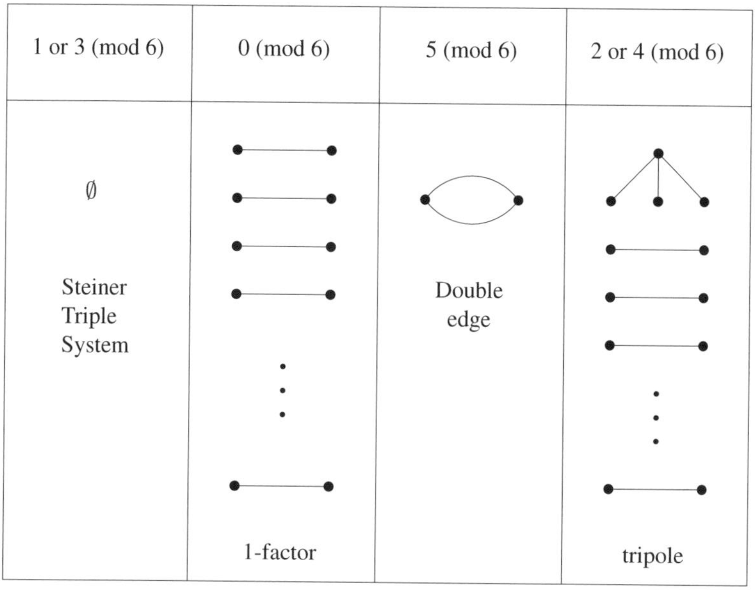

1 or 3 (mod 6)	0 (mod 6)	5 (mod 6)	2 or 4 (mod 6)
∅ Steiner Triple System	1-factor	Double edge	tripole

Figure 3.4: *Paddings of minimum coverings.*

The examples in Example 3.1.2 illustrate the three possibilities for $v \equiv 0, 2, 4$, or 5 (mod 6): the padding in (a) is a double edge, in (b) is a 1-factor, and in (c) is a tripole.

Exercises

3.1.3 Verify that if the leave of a packing of order $v \equiv 0$ or 2 (mod 6) is a 1-factor, then the packing is a MPT. (Hint: Each vertex belongs to an *odd* number $(= v - 1)$ of edges, and the edges are used up 2 at a time at each vertex w by the triangles in T containing w.)

3.1.4 Verify that if the leave of a packing of order $v \equiv 4$ (mod 6) is a tripole, then the packing is a MPT.

3.1.5 Verify that if the leave of a packing of order $v \equiv 5$ (mod 6) is a 4-cycle, then the packing is a MPT. (Hint: Each vertex belongs to an *even* number of edges, and the edges are used up 2 at a time at each vertex w by the triangles in T containing w.)

3.1.6 Verify that if the padding of a covering of order $v \equiv 0$ (mod 6) is a 1-factor, then the covering is a MCT.

3.1.7 Verify that if the padding of a covering of order $v \equiv 2$ or 4 (mod 6) is a tripole, then the covering is a MCT. (Hint: Each vertex belongs to an odd number $(= v - 1)$ of edges which are used up 2 at a time at each vertex w by the triangles in T containing w. Hence the padding must be a spanning graph with each vertex having odd degree.)

3.1.8 Verify that if the padding of a covering of order $v \equiv 5$ (mod 6) is a double edge, then the covering is a MCT.

3.2 Maximum packings.

We will break the constructions for maximum packings into four parts depending (not too surprisingly) on the leave. Of course, if $L = \emptyset$, then a maximum packing is simply a Steiner triple system. (We already know lots of ways to construct these!) So, we need only give constructions when the leave is a 1-factor, a 4-cycle, or a tripole.

The 1-factor Construction. Let $v \equiv 0$ or 2 (mod 6) and let (S, T) be a Steiner triple system of order $|S| = v + 1 \equiv 1$ or 3 (mod 6). Let $\infty \in S$ and let $T(\infty)$ be the set of *all* triples containing ∞. Let $L = \{\{a, b\} \mid \{\infty, a, b\} \in T\}$. Then $(S\backslash\{\infty\}, T\backslash T(\infty), L)$ is a maximum packing of order v with leave the 1-factor L.

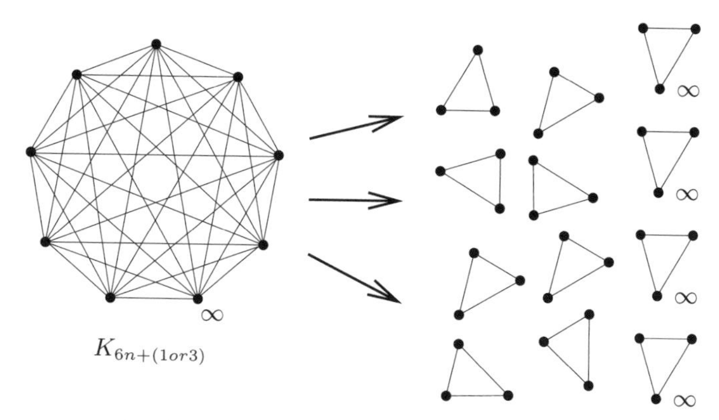

Steiner Triple System of order $6n+1$ or $6n+3$.

Delete ∞ and ALL edges containing ∞.

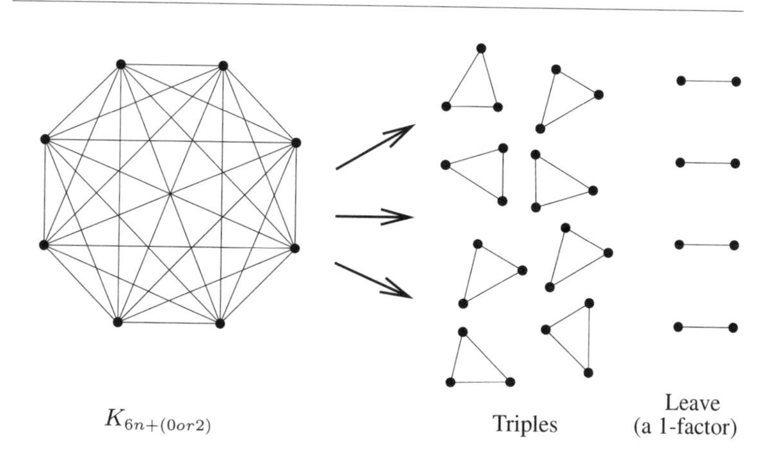

Figure 3.5: *Maximum packing of order* $6n + (0$ *or* $2)$ *with leave a* 1*-factor.*

Example 3.2.1 (MPT of order 6.)

$$S = \{\infty, 1, 2, 3, 4, 5, 6\}$$
$$T = \{\{\infty, 1, 3\}, \{\infty, 4, 5\}, \{\infty, 2, 6\}, \{1, 2, 4\}, \{2, 3, 5\}, \{1, 5, 6\}, \{3, 4, 6\}\}$$

Then $(S\backslash\{\infty\}, T\backslash T(\infty), L)$ is a maximum packing of order 6 with leave the 1-factor $L = \{\{1, 3\}, \{4, 5\}, \{2, 6\}\}$.

The 4-cycle Construction. Let $v \equiv 5 \pmod 6$ and let (P, B) be a PBD with one block of size 5 and the remaining blocks of size 3. Let T be the set of triples in B. Replace the block $\{a, b, c, d, e\}$ of size 5 with the maximum packing $(\{a, b, c, d, e\}, T^*, L)$, where $T^* = \{\{a, b, c\}, \{a, d, e\}\}$ and $L = \{\{b, e\},$

$\{e,c\},\{c,d\},\{d,b\}\}$. Then $(P,T\cup T^*,L)$ is a maximum packing of order v with leave the 4-cycle L.

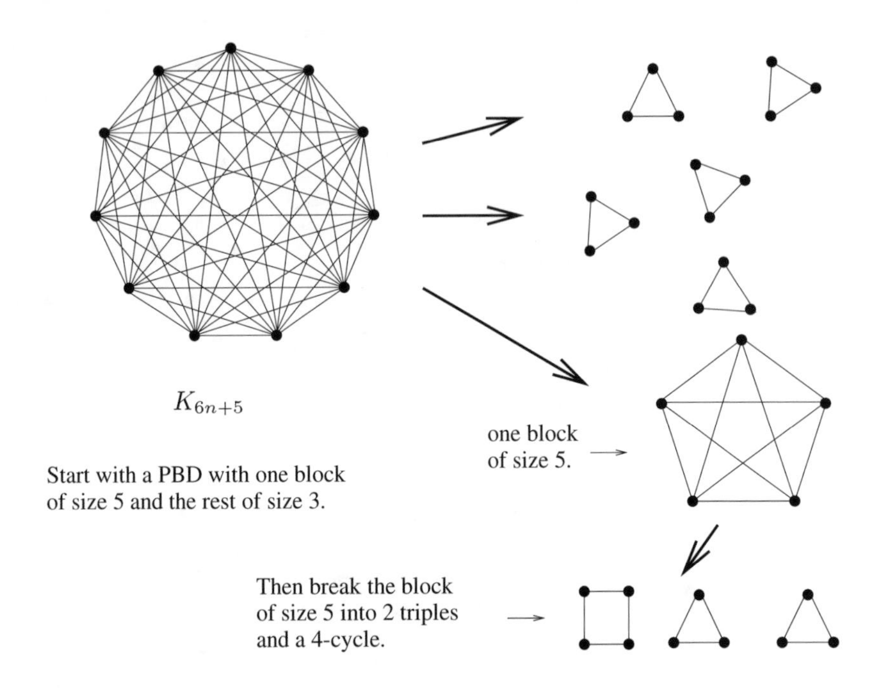

Figure 3.6: *Maximum packing on order* $6n+5$ *with leave a* 4*-cycle.*

Example 3.2.2 (MPT of order 11.)

$$P = \{1,2,3,4,5,6,7,8,9,10,11\}$$

$$B = \left\{ \begin{array}{llll} 1\,2\,9 & 1\,4\,7 & 2\,8\,10 & 1\,3\,5\,8\,11 \\ 3\,4\,9 & 1\,6\,10 & 3\,7\,10 & \\ 5\,6\,9 & 2\,3\,6 & 4\,5\,10 & \\ 7\,8\,9 & 2\,4\,11 & 4\,6\,8 & \\ 9\,10\,11 & 2\,5\,7 & 6\,7\,11 & \end{array} \right.$$

Then (P,T,L) is a maximum packing of order 11, where $T = (B\backslash \{1,3,5,8,11\}) \cup \{\{1,3,5\},\{1,8,11\}\}$, and with leave the 4-cycle $L=\{\{3,8\},\{8,5\},\{5,11\},\{11,3\}\}$.

The Tripole Construction. Let $v \equiv 4 \pmod 6$ and let (P,B) be a PBD of order $v+1$ with one block $\{\infty,a,b,c,d\}$ of size 5 and the remaining blocks of size 3. Denote by $B(\infty)$ the set of blocks containing ∞, and by L the set $\{\{x,y\} \mid \{x,y,\infty\} \in B\} \cup \{\{a,b\},\{a,c\},\{a,d\}\}$. (Note that L is a tripole.) Then $(P\backslash\{\infty\},(B\backslash B(\infty)) \cup \{\{b,c,d\}\},L)$ is a maximum packing of order v with leave a tripole.

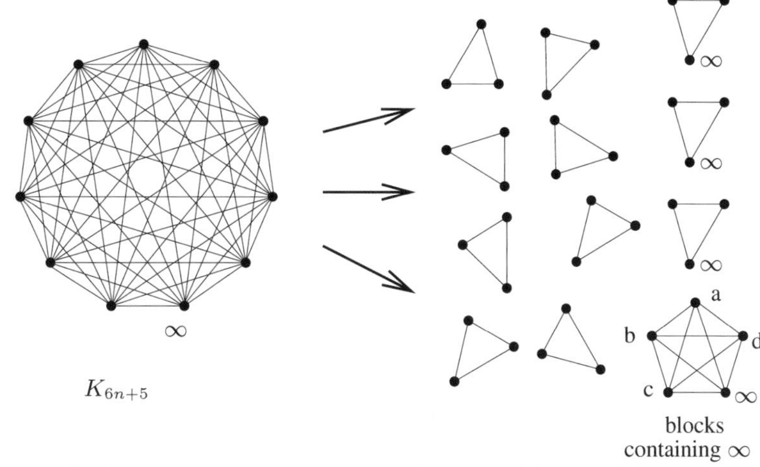

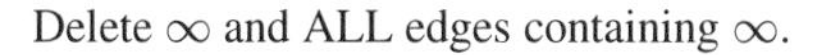

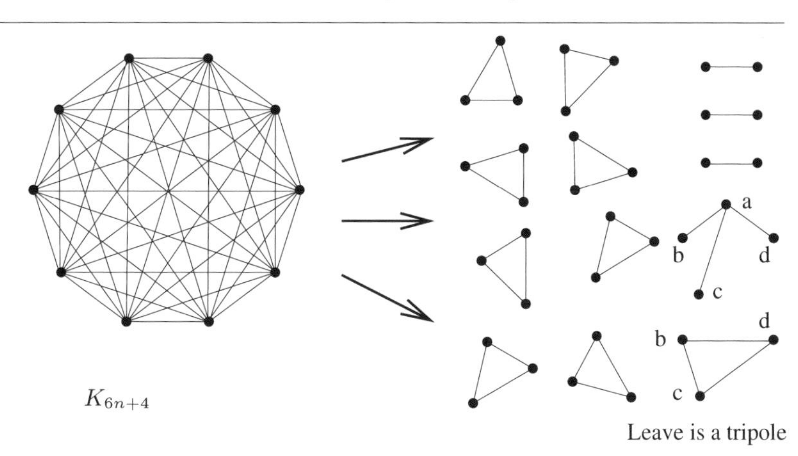

Figure 3.7: *Maximum packing of order* $6n + 4$ *with leave a tripole.*

Example 3.2.3 (MPT of order 10**.)**

$$P = \{\infty, 1, 2, 3, 4, 5, 6, 7, 8, 9, 10\}$$

$$B = \left\{ \begin{array}{llll} 1\,2\,9 & 1\,4\,7 & 2\,8\,10 & 1\,3\,5\,8\,\infty \\ 3\,4\,9 & 1\,6\,10 & 3\,7\,10 & \\ 5\,6\,9 & 2\,3\,6 & 4\,5\,10 & \\ 7\,8\,9 & 2\,4\,\infty & 4\,6\,8 & \\ 9\,10\,\infty & 2\,5\,7 & 6\,7\,\infty & \end{array} \right.$$

$$B(\infty) = \{\{9, 10, \infty\}, \{2, 4, \infty\}, \{6, 7, \infty\}, \{1, 3, 5, 8, \infty\}\}$$

Then $(P \setminus \{\infty\}, (B \setminus b(\infty)) \cup \{\{1,3,5\}\}, L)$ is a maximum packing of order 10, with leave the tripole $L = \{\{8,1\},\{8,3\},\{8,5\},\{2,4\},\{6,7\},\{9,10\}\}$.

Exercises

3.2.4 Let (S,T,L) be a triple where $|S| = v$, and T is a collection of triples T, and L is a collection of edges (T and L are based on S). Show that if every edge is in L or belongs to a triple in T and $3|T| + |L| \leq \binom{v}{2}$, then (S,T,L) is a *packing*.

3.2.5 Show that if (S,T,L) is a packing and L is a 1-factor, then $v \equiv 0$ or 2 (mod 6) and therefore (S,T,L) is a maximum packing. (Compare with Exercise 3.1.3.)

3.2.6 Show that if (S,T,L) is a packing and L is a 4-cycle, then $v \equiv 5$ (mod 6) and therefore (S,T,L) is a maximum packing. (Compare with Exercise 3.1.5.)

3.2.7 Show that if (S,T,L) is a packing and L is a tripole, then $v \equiv 4$ (mod 6) and therefore (S,T,L) is a maximum packing. (Compare with Exercise 3.1.4.)

3.2.8 Let $(Q, \circ)$ be the following quasigroup of order 7.

(Q, ∘) =

∘	1	2	3	4	5	6	7
1	1	6	5	7	2	3	4
2	6	2	7	5	4	1	3
3	5	7	3	6	1	4	2
4	7	5	6	4	3	2	1
5	2	4	1	3	5	7	6
6	3	1	4	2	7	6	5
7	4	3	2	1	6	5	7

This quasigroup and the Bose Construction can be used to construct a STS(21) on the symbols $\{1,\ldots,7\} \times \{1,2,3\}$. This STS(21) can be used to construct a MPT(20) by removing the symbol $(1,1)$. In this MPT(2), decide if the following pairs are in the leave or a triple, and if it is in a triple then find the triple.

(i) $(2,1)$ and $(4,1)$
(ii) $(2,1)$ and $(4,2)$
(iii) $(2,1)$ and $(6,2)$
(iv) $(1,2)$ and $(1,3)$.

3.2.9 The quasigroup $(Q, \circ)$ in Exercise 3.2.8 and the $6n + 5$ Construction can be used to construct a PBD(23) on the symbols $\{\infty_1, \infty_2\} \cup (\{1, \ldots, 7\} \times \{1, 2, 3\})$ with one block $\{\infty_1, \infty_2, (1, 1), (1, 2), (1, 3)\}$ of size 5 and the rest of size 3. This PBD(23) can be used to construct a MPT(23) by replacing the block of size 5 with the 4-cycle $\{\{\infty_1, (1, 2)\}, \{(1, 2), \infty_2\}, \{\infty_2, (1, 3)\}, \{(1, 3), \infty_1\}\}$ and the two triples $\{(1, 1), (1, 2), (1, 3)\}$ and $\{\infty_1, \infty_2, (1, 1)\}$. In this MPT(23), decide if the following pairs are in the leave or are in a triple, and if it is in a triple then find the triple.

(i) ∞_1 and $(1, 1)$
(ii) ∞_1 and $(1, 2)$
(iii) ∞_1 and $(2, 1)$
(iv) $(2, 1)$ and $(4, 1)$.

3.2.10 The PBD(23) constructed Exercise 3.2.9 above can be used to construct a MPT(22) by deleting the point ∞_2 and by letting the resulting tripole contain the edges $\{\infty_1, (1, 1)\}, \{\infty_1, (1, 2)\}$ and $\{\infty_1, (1, 3)\}$. In this MPT(22) decide if the following pairs are in the leave or a triple, and if in a triple then find the triple.

(i) ∞_1 and $(1, 3)$
(ii) ∞_1 and $(2, 1)$
(iii) $(3, 1)$ and $(3, 2)$
(iv) $(3, 1)$ and $(3, 3)$
(v) $(1, 1)$ and $(1, 2)$
(vi) $(3, 2)$ and $(4, 2)$.

3.3 Minimum coverings.

Just as was the case with maximum packings we will break the constructions into four parts depending on the padding. When $v \equiv 1$ or $3 \pmod 6$, the padding is the empty set so there is nothing to do. As with maximum packing we need only consider the cases where the padding is a 1-factor, a tripole, or a double edge. We will give each construction in turn with an example, then give pictorial representations of the constructions in Figures 3.8, 3.9, and 3.10.

The 1-factor covering Construction. Let $v \equiv 0 \pmod 6$ and let (X, B) be a PBD of order $v - 1 \equiv 5 \pmod 6$ with one block $\{a, b, c, d, e\}$ of size 5 and the remaining blocks of size 3. Denote by T the collection of blocks of size 3. Let $S = \{\infty\} \cup X$ and let $\pi = \{\{x_1, y_1\}, \{x_2, y_2\}, \ldots, \{x_t, y_t\}\}$ be any partition of $X \backslash \{a, b, c, d, e\}$. Let $\pi(\infty) = \{\{\infty, x_1, y_1\}, \{\infty, x_2, y_2\}, \ldots, \{\infty, x_t, y_t\}\}$ and $F(\infty) = \{\{\infty, a, e\}, \{\infty, b, e\}, \{\infty, c, d\}, \{c, d, e\}, \{a, b, d\}, \{a, b, c\}\}$.

(Note that $F(\infty)$ is an MCT of order 6.) Then (S, T^*, P) is a minimum covering of order v, where $T^* = T \cup \pi(\infty) \cup F(\infty)$, and $P = \pi \cup \{\{a, b\}, \{c, d\}, \{e, \infty\}\}$. (See Figure 3.8.)

Example 3.3.1 (MCT of order 12.)

$$X = \{1, 2, 3, 4, 5, 6, 7, 8, 9, 10, 11\}$$

$$B = \begin{array}{llll} 1\,2\,9 & 1\,4\,7 & 2\,8\,10 & 1\,3\,5\,8\,11 \\ 3\,4\,9 & 1\,6\,10 & 3\,7\,10 & \\ 5\,6\,9 & 2\,3\,6 & 4\,5\,10 & \\ 7\,8\,9 & 2\,4\,11 & 4\,6\,8 & \\ 9\,10\,11 & 2\,5\,7 & 6\,7\,11 & \end{array}$$

$F(\infty) = \{\{\infty, 1, 11\}, \{\infty, 3, 11\}, \{\infty, 5, 8\}, \{3, 5, 8\}, \{1, 3, 8\}, \{1, 3, 5\}\}\}$. Let $\pi = \{\{2, 4\}, \{6, 7\}, \{9, 10\}\}$ and $S = \{\infty\} \cup X$. Then (S, T^*, P) is a minimum covering of order 12 with padding the 1-factor $P = \pi \cup \{\{1, 3\}, \{5, 8\}, \{\infty, 11\}\}$.

The double edge covering Construction. Let $v \equiv 5 \pmod 6$ and let (S, B) be a PBD of order n with one block $\{a, b, c, d, e\}$ of size 5 and the remaining blocks of size 3. Denote by T the collection of blocks of size 3, and let $T^* = \{\{a, b, c\}, \{a, b, d\}, \{a, b, e\}, \{c, d, e\}\}$. Then $(S, T \cup T^*, P)$ is a minimum covering of order v, where $P = \{\{a, b\}, \{a, b\}\}$. (See Figure 3.9.)

Example 3.3.2 (MCT of order 11.) In Example 3.3.1 replace the block $\{1, 3, 5, 8, 11\}$ with the triples $T^* = \{\{1, 3, 5\}, \{1, 3, 8\}, \{1, 3, 11\}, \{5, 8, 11\}\}$. Then $(X, B \backslash \{\{1, 3, 5, 8, 11\}\} \cup T^*, P)$ is a minimum covering of order 11 with padding the double edge $P = \{\{1, 3\}, \{1, 3\}\}$.

The tripole covering Construction. Let $v \equiv 2$ or $4 \pmod 6$ and let (X, T) be a Steiner triple system of order $v - 1 \equiv 1$ or $3 \pmod 6$. Let $\{a, b, c\} \in T$ and let $\pi = \{\{x_1, y_1\}, \{x_2, y_2\}, \ldots, \{x_t, y_t\}\}$ be any partition of $X \backslash \{\{a, b, c\}\}$. Let $\pi(\infty) = \{\{\infty, x_1, y_1\}, \{\infty, x_2, y_2\}, \ldots, \{\infty, x_t, y_t\}\}$ and $T(\infty) = \{\{\infty, a, b\}, \{\infty, b, c\}, \{a, b, c\}\}$. Let $S = \{\infty\} \cup X$. Then (S, T^*, P) is a minimum covering of order v, where $T^* = T \cup \pi(\infty) \cup T(\infty)$, and $P = \pi \cup \{\{a, b\}, \{\infty, b\}, \{b, c\}\}$. (See Figure 3.10).

Example 3.3.3 (MCT of order 8.) Let (X, T) be the Steiner triple system of order 7 where $X = \{1, 2, 3, 4, 5, 6, 7\}$ and $T = \{\{1, 2, 4\}, \{2, 3, 5\}, \{3, 4, 6\}, \{4, 5, 7\}, \{5, 6, 1\}, \{6, 7, 2\}, \{7, 1, 3\}\}$. Let $\pi = \{\{3, 5\}, \{6, 7\}\}$ be a partition of $X \backslash \{\{1, 2, 4\}\}$, $\pi(\infty) = \{\{\infty, 3, 5\}, \{\infty, 6, 7\}\}$, and $T(\infty) = \{\{\infty, 1, 2\}, \{\infty, 2, 4\}, \{1, 2, 4\}\}$. Then (S, T^*, P) is a minimum covering of order 8, where $T^* = T \cup \pi(\infty) \cup T(\infty)$, and P is the tripole $\{\{3, 5\}, \{6, 7\}, \{2, 1\}, \{2, 4\}, \{2, \infty\}\}$.

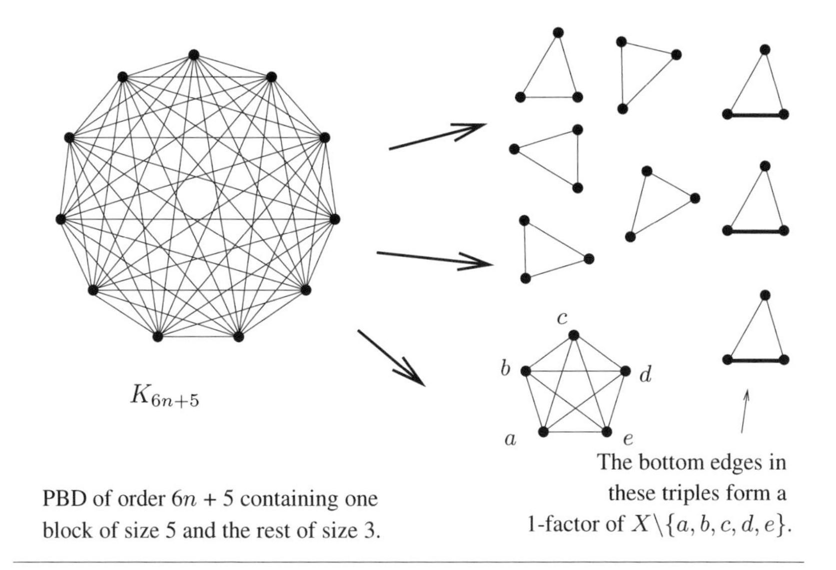

Add the point ∞ to K_{6n+5}.

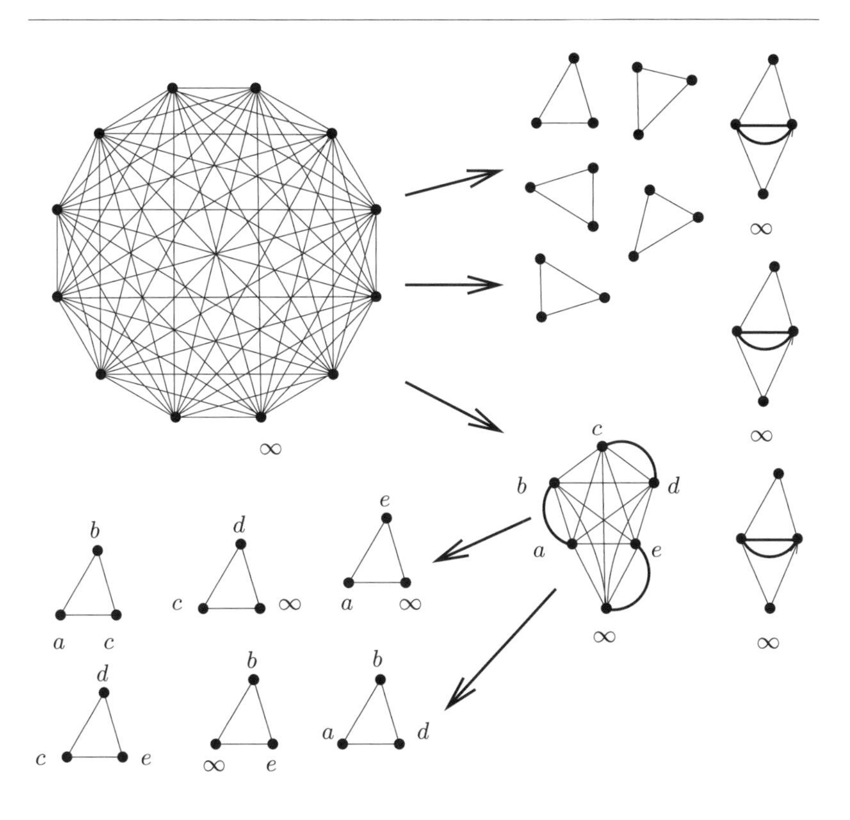

Figure 3.8: *Minimum covering of order $v \equiv 0 \pmod 6$ with padding a 1-factor.*

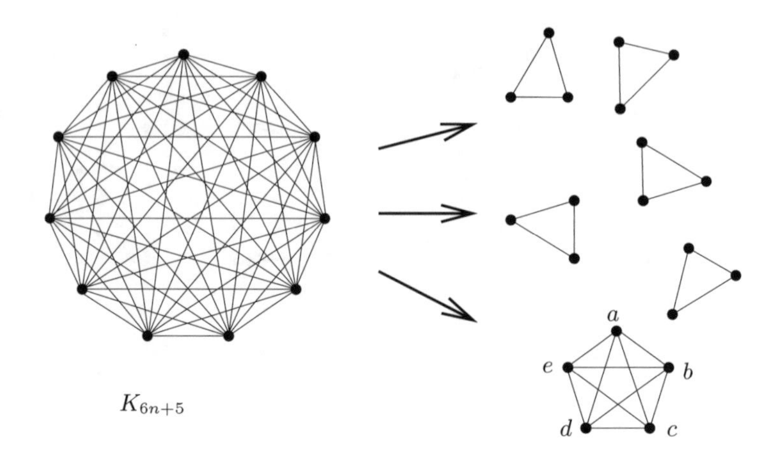

PBD of order $6n + 5$ containing one block of size 5 and the rest of size 3.

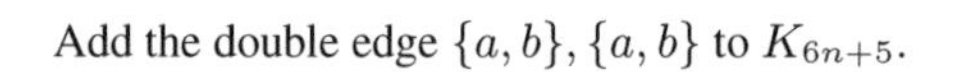
Add the double edge $\{a, b\}, \{a, b\}$ to K_{6n+5}.

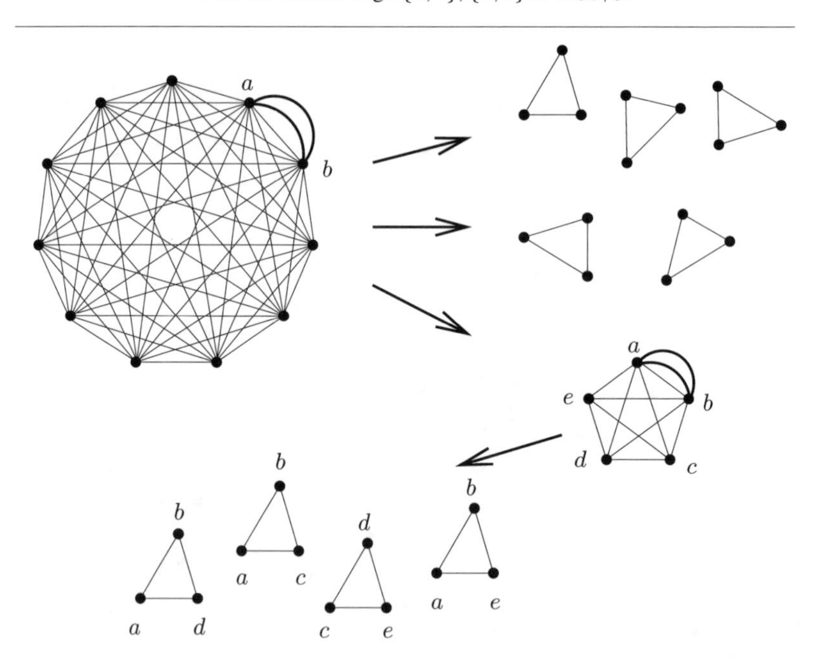

Figure 3.9: *Minimum covering of order* $v \equiv 5$ *(mod 6) with padding a double edge.*

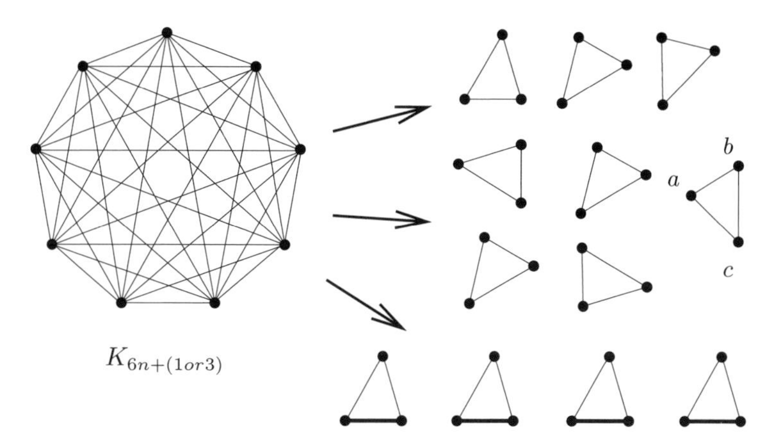

The bottom edges form a 1-factor of $X \backslash \{a, b, c\}$

Add the point ∞ to $K_{6n+(1or3)}$

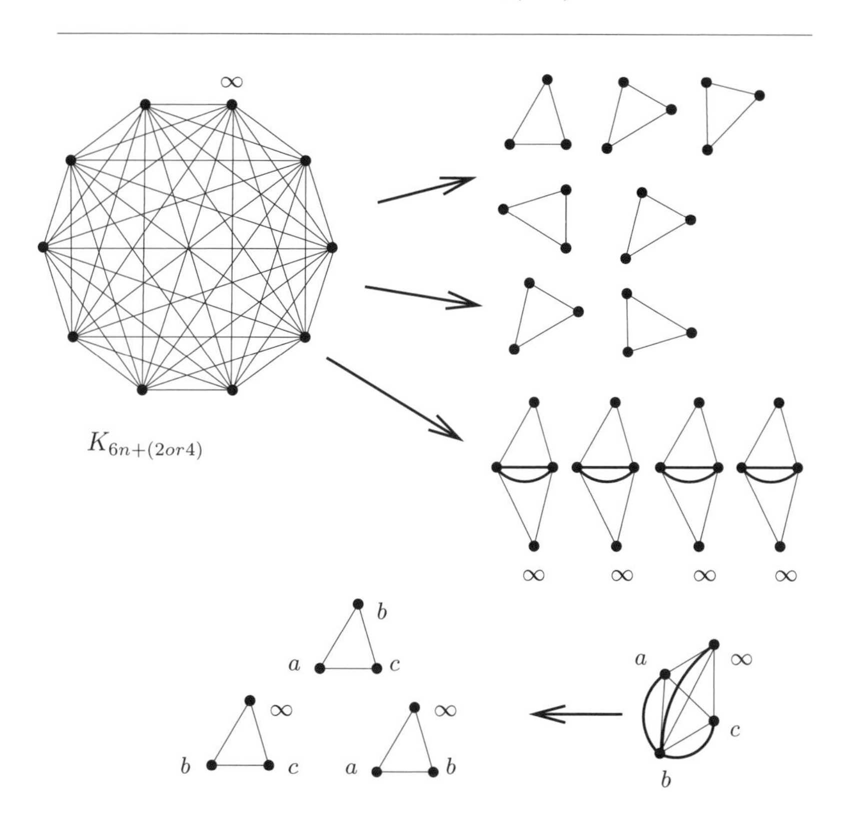

Figure 3.10: *Minimum covering of order $v \equiv 2$ or 4 (mod 6) with padding a tripole.*

Exercises

3.3.4 Let (S, T, P) be a triple where $|S| = v$, $P \subseteq E(\lambda K_v)$, and T is a collection of triples. Show that if each edge of $E(K_v) + P$ is in at least one triple of T and if $|T| \leq \binom{v}{2} + |P|$, then (S, T, P) is a *covering*.

3.3.5 Show that if (S, T, P) is a covering and P is a 1-factor, then $v \equiv 0$ (mod 6) and therefore (S, T, P) is a minimum covering. (Compare with Exercise 3.1.6.)

3.3.6 Show that if (S, T, P) is a covering and P is a double edge, then $v \equiv 5$ (mod 6) and therefore (S, T, P) is a minimum covering. (Compare with Exercise 3.1.8.)

3.3.7 Show that if (S, T, P) is a covering and P is a tripole, then $v \equiv 2$ or 4 (mod 6) and therefore (S, T, P) is a minimum covering. (Compare with Exercise 3.1.7.)

3.3.8 Let $(Q, \circ)$ be the following quasigroup of order 7.

(Q, ∘) =

∘	1	2	3	4	5	6	7
1	1	6	5	7	2	3	4
2	6	2	7	5	4	1	3
3	5	7	3	6	1	4	2
4	7	5	6	4	3	2	1
5	2	4	1	3	5	7	6
6	3	1	4	2	7	6	5
7	4	3	2	1	6	5	7

The quasigroup $(Q, \circ)$ and The $6n+5$ Construction can be used to construct a PBD(23) on the symbols $\{\infty_1, \infty_2\} \cup (\{1, \ldots, 7\} \times \{1, 2, 3\})$ with one block $\{\infty_1, \infty_2, (1,1), (1,2), (1,3)\}$ of size 5 and the rest of size 3. (See also Exercise 3.2.9.) This PBD(23) and the 1-factor covering construction can be used to construct an MCT(24) on the symbols $\{\infty, \infty_1, \infty_2\} \cup (\{1, \ldots, 7\} \times \{1, 2, 3\})$ by letting $\pi = \{\{(2, i), (3, i)\}, \{(4, i), (5, i)\}, \{(6, i), (7, i)\} \mid 1 \leq i \leq 3\}$ and letting a, b, c, d, e in the construction be defined by $a = \infty_1, b = \infty_2, c = (1,1), d = (1,2)$ and $e = (1,3)$. In this MCT(24), find all the triples containing the following pairs.

(i) ∞ and ∞_2

(ii) ∞ and $(1,3)$

(iii) ∞ and $(6,2)$

(iv) $(4, 1)$ and $(6, 1)$
(v) $(4, 1)$ and $(5, 1)$.

3.3.9 The PBD(23) constructed in Exercise 3.3.8 above and the double edge Construction can be used to construct a MCT(23), defining T^* by letting $a = \infty_1$, $b = \infty_2$, $c = (1, 1)$, $d = (1, 2)$ and $e = (1, 3)$. Find all the triples containing the following pairs.

(i) ∞_1 and ∞_2
(ii) ∞_1 and $(1, 1)$
(iii) ∞_1 and $(4, 1)$
(iv) $(4, 1)$ and $(5, 1)$
(v) $(4, 1)$ and $(6, 2)$.

3.3.10 The quasigroup $(Q, \circ)$ defined in Exercise 3.3.8 and the Bose Construction can be used to construct a STS(21) on the symbols $\{1, \ldots, 7\} \times \{1, 2, 3\}$. This STS(21) and the tripole covering Construction can be used to construct a MCT(22) on the symbols $\{\infty\} \cup (\{1, \ldots, 7\} \times \{1, 2, 3\})$ by letting $a = (1, 1)$, $b = (1, 2)$ and $c = (1, 3)$ (recall that $\{(1, 1), (1, 2), (1, 3)\}$ is a triple in this STS(21)), and letting $\pi = \{\{(2, i), (3, i)\}, \{(4, i), (5, i)\}, \{(6, i), (7, i)\} \mid 1 \leq i \leq 3\}$. Find all the triples containing the following pairs.

(i) ∞ and $(1, 1)$
(ii) ∞ and $(1, 2)$
(iii) ∞ and $(3, 3)$
(iv) $(1, 1)$ and $(1, 2)$
(v) $(3, 1)$ and $(5, 1)$
(vi) $(3, 1)$ and $(4, 1)$.

Kirkman Triple Systems

4.1 A recursive construction

In this chapter we will consider Steiner triple systems with the additional property of being resolvable. Not only will we consider when such systems exist, but also many different PBDs will be found, and in the process several important constructions for PBDs will be introduced. In fact, understanding the various constructions in Section 2.2 is more to the point than finding the PBDs themselves.

We begin with some definitions. A *parallel class* in a Steiner triple system (S, T) is a set of triples in T that partitions S. A STS (S, T) is *resolvable* if the triples in T can be partitioned into parallel classes. A resolvable $STS(v)$ is also known as a *Kirkman triple system* of order v, or $KTS(v)$.

Example 4.1.1 A $KTS(9)$ with parallel classes π_1, π_2, π_3 and π_4 can be defined as follows:

π_1			π_2			π_3			π_4		
1	2	3	1	4	7	1	5	9	1	6	8
4	5	6	2	5	8	2	6	7	2	4	9
7	8	9	3	6	9	3	4	8	3	5	7

A $KTS(15)$ with parallel classes $\pi_1, \pi_2, \pi_3, \pi_4, \pi_5, \pi_6$ and π_7 can be defined as follows:

π_1			π_2			π_3			π_4		
1	2	3	1	4	5	1	6	7	1	8	9
4	8	12	2	8	10	2	9	11	2	12	15
5	10	14	3	13	15	3	12	14	3	5	6
6	11	13	6	9	14	4	10	15	4	11	14
7	9	15	7	11	12	5	8	13	7	10	13

π_5			π_6			π_7		
1	10	11	1	12	13	1	14	15
2	13	14	2	4	6	2	5	7
3	4	7	3	9	10	3	8	11
5	9	12	5	11	15	4	9	13
6	8	15	7	8	14	6	10	12

Exercises

4.1.2 Show that in any $KTS(v)$,

(a) the number of triples in each parallel class is $v/3$, and
(b) the number of parallel classes is $(v-1)/2$.

Deduce that $v \equiv 3 \pmod 6$ is necessary.

Kirkman triple systems are named after the Rev. T. P. Kirkman who in 1850 posed and solved [13, 14] the "Kirkman schoolgirl problem": Is it possible for a schoolmistress to take 15 schoolgirls on a walk each day of the 7 days of a week, walking with 5 rows of 3 girls each, in such a way that each pair of girls walks together in the same row on exactly one day? A little reflection shows that this problem is asking if there exists a $KTS(15)$. The answer is clearly yes, since a $KTS(15)$ is exhibited in Example 4.1.1 (you can think of $\pi_1, \pi_2, \pi_3, \pi_4, \pi_5, \pi_6,$ π_7 as the arrangement of the girls walking on Monday through Sunday.

Recall that a *pairwise balanced design* of order v (or simply, a $PBD(v)$) is an ordered pair (S, B) where S is a finite set of symbols with $|S| = v$, and B is a set of subsets of S called *blocks* (not necessarily of the same size), such that each pair of distinct elements of S occurs together in exactly one block in B. So a Steiner triple system is a PBD with all blocks of size 3. The following recursive construction shows how a $PBD(3n+1)$ can be used to construct a $KTS(6n+3)$ (notice that the proof actually constructs such a KTS).

Theorem 4.1.3 *If there exists a $PBD(3n+1)$ with block sizes $k_1, k_2, \ldots, k_x$, and if there exists a $KTS(2k_i+1)$ for $1 \leq i \leq x$, then there exists a $KTS(6n+3)$.*

Proof Let (P, B) be a $PBD(3n+1)$, with $P = \{1, 2, \ldots, 3n+1\}$ and with block sizes $k_1, k_2, \ldots, k_x$. Define a $KTS(6n+3)$ (S, T) with $S = \{\infty\} \cup (\{1, 2, \ldots, 3n+1\} \times \{1, 2\})$ as follows.

(1) For $1 \leq i \leq 3n+1$, $\{\infty, (i,1), (i,2)\} \in T$, and

(2) for each block $b \in B$, let $(S(b), T(b))$ be a KTS of order $2|b|+1$, where $S(b) = \{\infty\} \cup (b \times \{1, 2\})$ and where the symbols have been named so that $\{\infty, (i,1), (i,2)\} \in T(b)$ for all $i \in b$, and let $T(b) \setminus \{\{\infty, (i,1), (i,2)\} \mid i \in b\} \subseteq T$.

The fact that this defines a $STS(6n+3)$ is left to Exercise 4.1.7 (see also Figure 2.1), but we now show how the triples in T can be partitioned into parallel classes $\pi_1, \pi_2, \ldots, \pi_{3n+1}$ (see Figure 4.2).

For $1 \leq x \leq 3n+1$, let B_x be the set of blocks in B that contains symbol x. For each block $b \in B_x$, let $p_x(b)$ be the parallel class of triples in $T(b)$ that contains the triple $\{\infty, (x,1), (x,2)\}$ (recall that in (2) the symbols are named so that $T(b)$ does contain this triple). Then define

$$\pi_x = \bigcup_{b \in B_x} p_x(b)$$

where, of course, the triple $\{\infty, (x,1), (x,2)\}$ is listed just once. □

This construction is exhibited graphically in Figures 4.1 and 4.2.

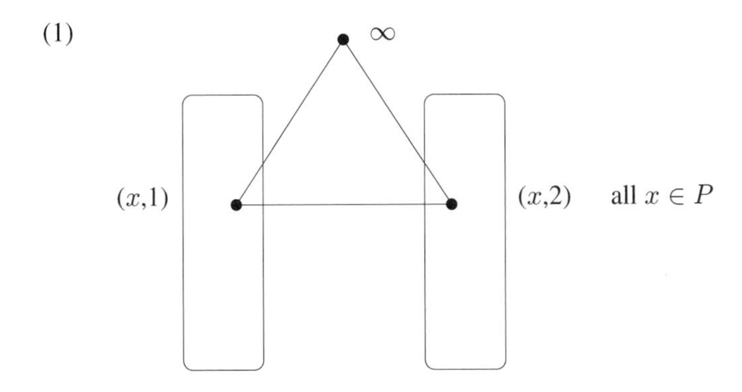

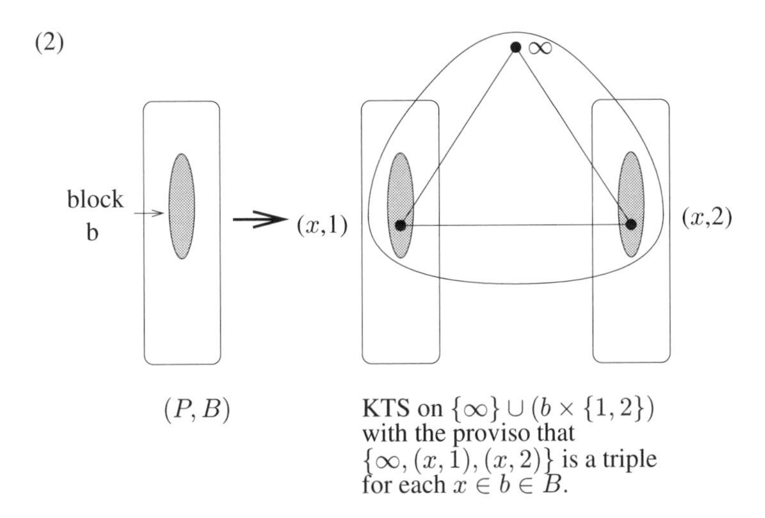

Figure 4.1: *Kirkman triple system Construction.*

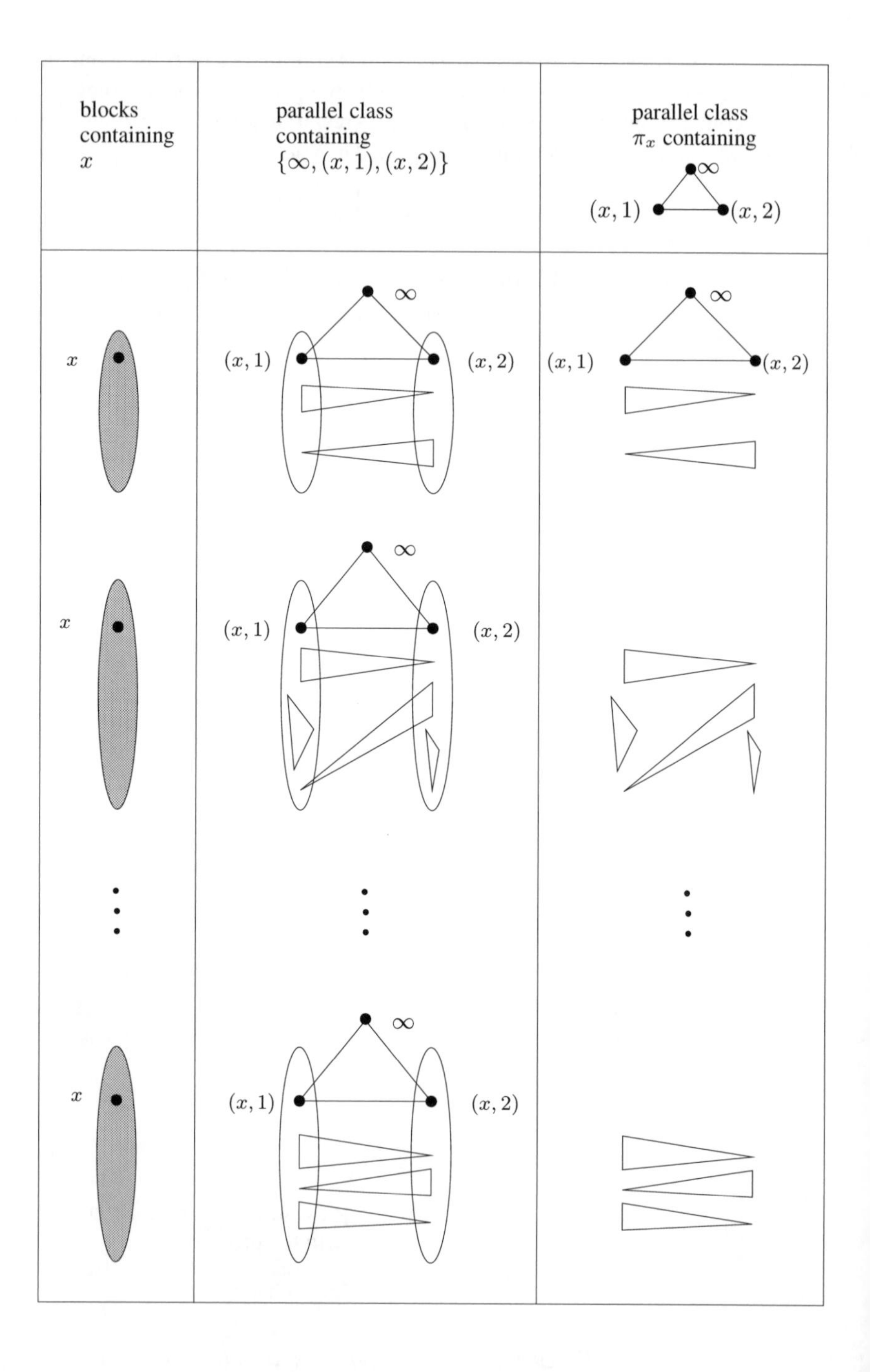

Figure 4.2: *The parallel class π_x.*

Example 4.1.4 This example looks at the construction in the proof of Theorem 4.1.3 in great detail.

Define a PBD (P, B) of order $3n + 1 = 13$ in which all blocks have size 4 by setting $P = \{1, 2, \ldots, 13\}$ and $B = \{\{i, 1 + i, 3 + i, 9 + i\} | 1 \leq i \leq 13\}$, reducing all sums modulo 13. Since a KTS of order $2 \cdot 4 + 1 = 9$ is exhibited in Example 4.1.1, we can apply Theorem 4.1.3 to construct a KTS $(\{\infty\} \cup (\{1, \ldots, 13\} \times \{1, 2\}), T)$ of order 27 as follows.

(1) For $1 \leq i \leq 13$, $\{\infty, (i, 1), (i, 2)\} \in T$.

(2) We consider each block $b \in B$ in turn. Remember that the names of the symbols in the $KTS(9)$ $(\infty \cup (b \times \{1, 2\}), T(b))$ are chosen so that for each $i \in b$, $\{\infty, (i, 1), (i, 2)\} \in T(b)$. To this end, if $b = \{i, j, k, \ell\} \in B$, then we choose $i < j < k < \ell$ and rename the symbols $1, 2, \ldots, 9$ in the $KTS(9)$ in Example 4.1.1 with $\infty, (i, 1), (i, 2), (j, 1), (k, 1), (\ell, 1), (j, 2), (\ell, 2)$, and $(k, 2)$ respectively (of course, other choices for their renaming are possible). So, for example, for $b = \{1, 2, 4, 10\} \in B$, $T(b)$ contains the triples:

$$
\begin{array}{ll}
\{\infty, (1, 1), (1, 2)\} & \{\infty, (2, 1), (2, 2)\} \\
\{(2, 1), (4, 1), (10, 1)\} & \{(1, 1), (4, 1), (10, 2)\} \\
\{(2, 2), (10, 2), (4, 2)\} & \{(1, 2), (10, 1), (4, 2)\} \\
\{\infty, (4, 1), (4, 2)\} & \{\infty, (10, 1), (10, 2)\} \\
\{(1, 1), (10, 1), (2, 2)\} & \{(1, 1), (2, 1), (4, 2)\} \\
\{(1, 2), (2, 1), (10, 2)\} & \{(1, 2), (4, 1), (2, 2)\}
\end{array}
$$

Similarly, for each $b = \{2, 3, 5, 11\}, \{3, 4, 6, 12\}, \ldots, \{1, 3, 9, 13\}$ in turn, 12 further triples are defined.

However, rather than continue writing out all the other triples, it is more instructive to consider how to find the triple containing a given pair, say $\{(y, i), (z, j)\}$. Of course, if $y = z$ then the triple is $\{\infty, (y, 1), (z, 2)\}$. If $y \neq z$, let $b \in B$ be the block containing y and z. Then (y, i) and (z, j) occur in $T(b)$. For example, we can find the triple containing $(5, 1)$ and $(7, 2)$ by observing that $\{5, 7\} \subseteq \{4, 5, 7, 13\} \in B$, and forming $T(b)$ as above reveals that $\{(4, 1), (5, 1), (7, 2)\} \in T(\{4, 5, 7, 13\})$.

We now consider the problem of finding the parallel classes. (See Figure 4.2.) To find, for example, the parallel class π_4, we begin by finding B_4, the set of all blocks in the PBD containing symbol 4:

$$B_4 = \{\{1, 2, 4, 10\}, \{3, 4, 6, 12\}, \{4, 5, 7, 13\}, \{4, 8, 9, 11\}\}.$$

For each block $b \in B_4$, we find $p_4(b)$, the parallel class in $T(b)$ that contains $\{\infty, (4, 1), (4, 2)\}$.

$$
\begin{aligned}
b_1 = \{1, 2, 4, 10\} : p_4(b_1) &= \{\{\infty, (4, 1), (4, 2)\}, \{(1, 1), (10, 1), (2, 2)\}, \\
&\quad \{(1, 2), (2, 1), (10, 2)\}\}.
\end{aligned}
$$

$$
\begin{aligned}
b_2 = \{3,4,6,12\} : p_4(b_2) &= \{\{\infty,(4,1),(4,2)\},\{(3,1),(6,1),(12,2)\},\\
&\quad \{(3,2),(12,1),(6,2)\}\}.\\
b_3 = \{4,5,7,13\} : p_4(b_3) &= \{\{\infty,(4,1),(4,2)\},\{(5,1),(7,1),(13,1)\},\\
&\quad \{(5,2),(13,2),(7,2)\}.\\
b_4 = \{4,8,9,11\} : p_4(b_4) &= \{\{\infty,(4,1),(4,2)\},\{(8,1),(9,1),(11,1)\},\\
&\quad \{(8,2),(9,2),(11,2)\}.
\end{aligned}
$$

Then $\pi_4 = p_4(b_1) \cup p_4(b_2) \cup p_4(b_3) \cup p_4(b_4)$, where of course, the triple $\{\infty,(4,1),(4,2)\}$ is listed exactly once.

Exercises

4.1.5 Continuing Example 4.1.4, find:

(a) π_6
(b) π_8
(c) π_{11}
(d) the triple in the $KTS(27)$ containing the pair of symbols

(i) (2, 1) and (7, 2)
(ii) (2, 1) and (8, 1)
(iii) (3, 1) and (3, 2)
(iv) (5, 2) and (6, 2)
(v) (5, 2) and ∞
(vi) $(5,1)$ and $(13,1)$
(vii) $(6,1)$ and $(8,1)$
(viii) ∞ and $(11,2)$
(ix) $(1,1)$ and $(13,2)$
(x) $(8,1)$ and $(8,2)$.

(e) the parallel class π_i containing the triple that contains the pair

(i) (2, 1) and (7, 2)
(ii) (2, 1) and (8, 1)
(iii) (3, 1) and (3, 2)
(iv) (5, 2) and (6, 2)
(v) (5, 2) and ∞
(vi) $(5,1)$ and $(13,1)$
(vii) $(6,1)$ and $(8,1)$
(viii) ∞ and $(11,2)$
(ix) $(1,1)$ and $(13,2)$
(x) $(8,1)$ and $(8,2)$.

4.1.6 Suppose there exists a PBD(28) that contains the blocks $\{1,2,12,26\}$, $\{3,7,8,12\}$, $\{4,5,12,13,14,19,20\}$, $\{6,12,21,24\}$, $\{9,12,15,16,$

$17, 18, 22\}$, $\{10, 11, 12, 23\}$ and $\{12, 25, 27, 28\}$. (You won't need to know any of the other blocks in this PBD to answer the questions below.) As in Theorem 4.1.3, this PBD can be used to construct a KTS $(57), (\{\infty\}\cup(\{1, 2, \ldots, 27\}\times\{1, 2\}), T)$. In this construction, use the $KTS(9)$ in Example 4.1.1 with the symbols $1, 2, \ldots, 9$ renamed with $\infty, (i, 1), (i, 2), (j, 1), (k, 1), (\ell, 1), (j, 2), (\ell, 2)$ and $(k, 2)$ respectively (as in (2) of Example 4.1.4), and for each block $b = \{i, j, k, \ell, a, b, c\}$ of size 7 use the $KTS(15)$ in Example 4.1.1 with the symbols $1, 2, \ldots,$ 15 renamed with $\infty, (i, 1), (i, 2), (j, 1), (j, 2), \ldots, (c, 1), (c, 2)$ respectively (this is one way to rename the symbols so that $\{\infty, (x, 1), (x, 2)\}$ is a triple for all $x \in b$ as (2) of Theorem 4.1.3 requires).

(a) Find the third symbol in the triple in the $KTS(57)$ that contains the pair of symbols

(i) $(3, 1)$ and $(8, 2)$
(ii) $(5, 1)$ and $(19, 1)$
(iii) $(15, 2)$ and $(22, 2)$
(iv) ∞ and $(12, 1)$.

(b) Find the value of x where π_x is the parallel class containing the triple that contains the pair

(i) $(13, 1)$ and $(14, 1)$
(ii) $(13, 1)$ and $(14, 2)$
(iii) $(13, 1)$ and $(13, 2)$
(iv) $(12, 1)$ and $(17, 1)$.

(c) Find the triple in the parallel class π_{12} that contains the symbol

(i) $(14, 2)$
(ii) $(20, 1)$
(iii) $(16, 1)$
(iv) $(26, 2)$
(v) $(7, 1)$
(vi) $(11, 1)$.

(d) Find *all* triples in the parallel class π_{12}.

4.1.7 Show that the construction of a $KTS(6n+3)$ in the proof of Theorem 4.1.3 does produce a STS.

We end this section by stating two results. The first result provides the right ingredients to use with the recursive construction of Theorem 4.1.3 to settle the existence problem for Kirkman triple systems, which is the second result.

Theorem 4.1.8 *For all $n \geq 1$, there exists a $PBD(3n + 1)$ with block sizes in $\{4, 7, 10, 19\}$.*

Proof This is left to Section 4.2. □

Theorem 4.1.9 *For all $n \geq 1$, there exists a $KTS(6n+3)$.*

Proof In view of Theorems 4.1.3 and 4.1.8, all we need to do is find a $KTS(2k+1)$ for each $k \in \{4, 7, 10, 19\}$. A $KTS(9)$ and a $KTS(15)$ are constructed in Example 4.1.1, so it remains to find a $KTS(21)$ and a $KTS(39)$.

The following 10 parallel classes define a $KTS(21)$:

	π_1			π_2			π_3			π_4	
1	2	3	4	5	6	7	8	9	10	11	12
4	7	13	7	10	16	10	13	19	1	13	16
5	8	14	8	11	17	11	14	20	12	14	20
6	9	15	9	12	18	12	15	21	3	15	18
10	17	21	3	13	20	2	6	16	5	9	19
11	18	19	1	14	21	3	4	17	6	7	20
12	16	20	2	15	19	1	5	18	4	8	21

	π_5			π_6			π_7			π_8	
13	14	15	16	17	18	19	20	21	10	18	20
4	16	19	1	7	19	1	4	10	2	13	21
5	17	20	2	8	20	2	5	11	3	5	16
6	18	21	3	9	21	3	6	12	6	8	19
1	8	12	4	11	15	7	14	18	1	9	11
2	9	10	5	12	13	8	15	16	4	12	14
3	7	11	6	10	14	9	13	17	7	15	17

	π_9			π_{10}	
11	16	21	12	17	19
3	14	19	1	15	20
1	6	17	2	4	18
4	9	20	5	7	21
2	7	12	3	8	10
5	10	15	6	11	13
8	13	18	9	14	16

The following 13 parallel classes π_i (for $1 \leq i \leq 13$):

$$\begin{aligned} \pi_i \;=\; & \{\{0+3i,\; 1+3i,\; 2+3i\}, \{3+3i,\; 9+3i,\; 27+3i\}, \\ & \{4+3i,\; 10+3i,\; 28+3i\}, \{5+3i,\; 11+3i,\; 29+3i\}, \\ & \{6+3i,\; 15+3i,\; 18+3i\}, \{7+3i,\; 16+3i,\; 19+3i\}, \\ & \{8+3i,\; 17+3i,\; 20+3i\}, \{12+3i,\; 31+3i,\; 38+3i\}, \\ & \{13+3i,\; 32+3i,\; 36+3i\}, \{14+3i,\; 30+3i,\; 37+3i\}, \\ & \{21+3i,\; 25+3i,\; 35+3i\}, \{22+3i,\; 26+3i,\; 33+3i\}, \\ & \{23+3i,\; 24+3i,\; 34+3i\}\}, \end{aligned}$$

together with the following 6 parallel classes:

$$\begin{aligned}
\pi_{14} &= \{\{12+3i, 32+3i, 37+3i\} \mid 0 \le i \le 12\},\\
\pi_{15} &= \{\{13+3i, 30+3i, 0+3i\} \mid 0 \le i \le 12\},\\
\pi_{16} &= \{\{14+3i, 31+3i, 36+3i\} \mid 0 \le i \le 12\},\\
\pi_{17} &= \{\{21+3i, 26+3i, 34+3i\} \mid 0 \le i \le 12\},\\
\pi_{18} &= \{\{22+3i, 24+3i, 35+3i\} \mid 0 \le i \le 12\}, \text{ and}\\
\pi_{19} &= \{\{23+3i, 25+3i, 33+3i\} \mid 0 \le i \le 12\},
\end{aligned}$$

reducing all sums modulo 39, defines a $KTS(39)$. □

4.2 Constructing pairwise balanced designs

In this section we will construct the pairwise balanced designs needed to prove Theorem 4.1.8. The point of this is really to introduce a variety of techniques that can be used to construct PBDs. It is extremely instructive to consider each PBD in turn. The constructions introduced in this fairly simple setting are used extensively in Chapter 5, so familiarity with the techniques developed here will aid in understanding later results.

We begin with the following lemma which constructs a lot of PBDs of small orders. These are then used to prove Theorem 4.1.8. The construction in Theorem 4.1.8 is inductive, so we begin by considering the cases where $4 \le v \le 46$ and $79 \le v \le 82$. (As stated above, the techniques used to construct these PBDs are of interest in their own right.)

Lemma 4.2.1 *For* $1 \le n \le 15$ *and for* $26 \le n \le 27$ *there exists a* $PBD(3n+1)$ *with block sizes in* $\{4, 7, 10, 19\}$.

Proof For each value of n in turn, we construct a $PBD(v)$ (P, B) with $v = 3n+1$, and block sizes in $\{4, 7, 10, 19\}$, as follows.
v = 4: B contains one block of size 4.
v = 7: B contains one block of size 7.
v = 10: B contains one block of size 10.
v = 13: One such PBD is constructed in Example 4.1.4 (this PBD is cyclic). We take the opportunity to exhibit another construction here that uses a technique which will be seen often in the following pages.

The PBD is defined with $P = \{\infty\} \cup (\{1, 2, 3\} \times \{1, 2, 3, 4\})$, and

$$\begin{aligned}
B &= \{\{(x,1), (y,2), (x \circ_1 y, 3), (x \circ_2 y, 4)\} | 1 \le x, y \le 3\}\}\\
&\quad \cup \{\{\infty, (1,\ell), (2,\ell), (3,\ell)\} | 1 \le \ell \le 4\},
\end{aligned}$$

where $(\{1,2,3\},\circ_1)$ and $(\{1,2,3\},\circ_2)$ are the quasigroups

$\circ_1$	1	2	3
1	1	3	2
2	3	2	1
3	2	1	3

and

$\circ_2$	1	2	3
1	1	2	3
2	3	1	2
3	2	3	1

(so all blocks have size 4). This construction is represented graphically in Figure 4.3.

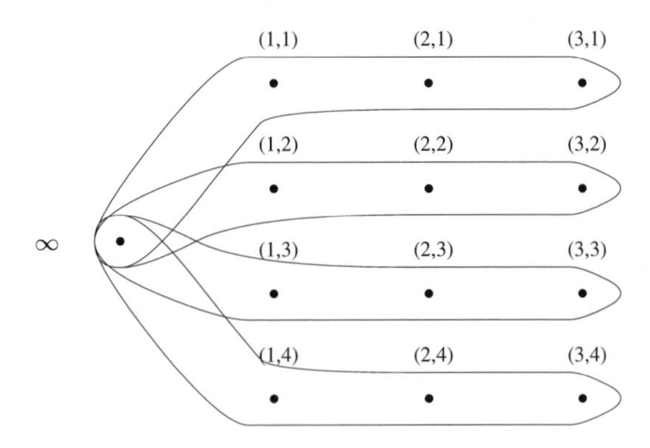

4 blocks of size 4.

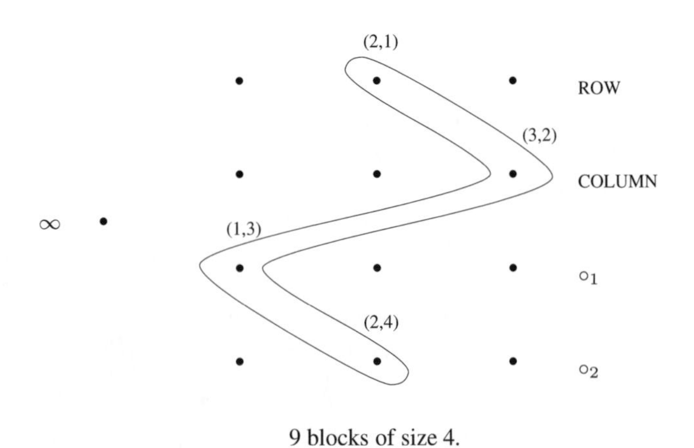

9 blocks of size 4.

Figure 4.3: *Using 2 MOLS(3) to construct a PBD(13).*

Of course, these two quasigroups are not just any quasigroups of order 3. Two latin squares L_1 and L_2 of order n are said to be *orthogonal* if for each $(i, j) \in \{1, 2, \ldots, n\} \times \{1, 2, \ldots, n\}$ there is exactly one ordered pair (x, y) such that cell (x, y) of L_1 contains the symbol i and cell (x, y) of L_2 contains the symbol j (or in quasigroup notation, $x \circ_1 y = i$ and $x \circ_2 y = j$ has a unique solution for x and y). In other words, when L_1 are L_2 are superimposed the resulting set of n^2 ordered pairs are *distinct*. These are also referred to as 2 *mutually orthogonal latin squares of order* n, or 2 MOLS(n). Two quasigroups are *orthogonal* if the corresponding latin squares are orthogonal. The two quasigroups of order 3 above are easily seen to be orthogonal. Why must the quasigroups used in this construction of the PBD(13) be orthogonal? Suppose we try to find the block of size 4 containing the symbols $(1, 3)$ and $(2, 4)$, for example. From the definition of B, we need to find x and y, where $x \circ_1 y = 1$ and $x \circ_2 y = 2$. It is orthogonality that ensures we get a unique solution for x and y. In this case, $x = 2$ and $y = 3$, so the required block is $\{(2, 1), (3, 2), (1, 3), (2, 4)\}$, the block depicted in Figure 4.3

v = 16: Again we construct a PBD in which all blocks have size 4, again using orthogonal quasigroups.

The following are two orthogonal quasigroups of order 4 (check it out!).

$\circ_1$	1	2	3	4
1	1	3	4	2
2	4	2	1	3
3	2	4	3	1
4	3	1	2	4

$\circ_2$	1	2	3	4
1	1	4	2	3
2	3	2	4	1
3	4	1	3	2
4	2	3	1	4

Define a $PBD(16)$ with $P = \{1, 2, 3, 4\} \times \{1, 2, 3, 4\}$ and

$$\begin{aligned} B &= \{\{(x, 1), (y, 2), (x \circ_1 y, 3), (x \circ_2 y, 4)\} | 1 \leq x, y \leq 4\} \\ & \quad \bigcup \{\{(1, z), (2, z), (3, z), (4, z)\} | 1 \leq z \leq 4\}. \end{aligned}$$

See Figure 4.4 for a pictorial representation of this construction.

v = 19: B contains one block of size 19.

v = 22: We start with the $KTS(15)$ in Example 4.1.1 with parallel classes $\pi_1, \pi_2, \ldots, \pi_7$. For $1 \leq i \leq 7$, add a new symbol $15 + i$ to each of the triples in π_i to produce blocks of size 4 on the symbols $1, \ldots, 22$ (so, since π_i is a parallel class, the symbol $15 + i$ occurs in a block of size 4 with each of the symbols 1 to 15). To these blocks of size 4 add the block $\{16, 17, 18, 19, 20, 21, 22\}$ of size 7 to produce the required $PBD(22)$.

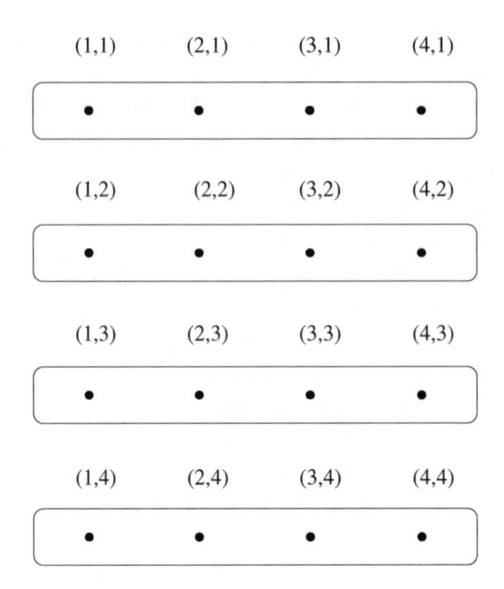

4 blocks of size 4.

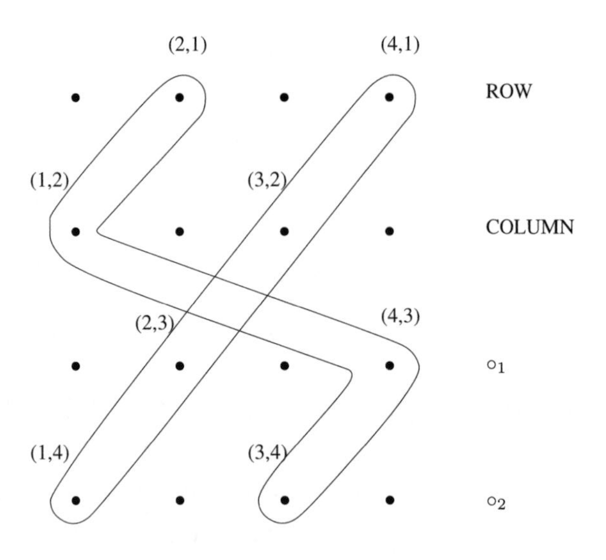

16 blocks of size 4.

Figure 4.4: *Using 2 MOLS(4) to construct a PBD(16).*

v = 25: In Section 1.7 we discussed cyclic Steiner triple systems, after introducing automorphisms of STSs. One advantage of a cyclic STS is that by defining just the base blocks, the entire set of blocks has been described. We will now construct a $PBD(25)$ in which all blocks have size 4. Unfortunately no such cyclic PBD exists, but we can find one which has a different automorphism (of course, an *automorphism* of a PBD (P, B) is a bijection $\alpha : P \to P$ such that for each $b = \{x_1, x_2, \ldots, x_k\} \in B$, $b\alpha = \{x_1\alpha, x_2\alpha, \ldots, x_k\alpha\} \in B$).

Set $P = \{1, 2, 3, 4, 5\} \times \{1, 2, 3, 4, 5\}$. It is convenient to introduce some notation: if $b = \{(c_1, d_1), (c_2, d_2), (c_3, d_3), (c_4, d_4)\}$, then let $b + (i, j) = \{(c_1 + i, d_1 + j), (c_2 + i, d_2 + j), (c_3 + i, d_3 + j), (c_4 + i, d_4 + j)\}$. Now define the base blocks

$$b_1 = \{(1,1), (1,2), (2,1), (3,3)\} \text{ and } b_2 = \{(1,1), (1,3), (3,1), (5,5)\},$$

and define $B = \{b_1 + (i, j), b_2 + (i, j) | 0 \leq i \leq 4, 0 \leq j \leq 4\}$, where all sums are reduced modulo 5. Then (P, B) is a $PBD(25)$ which clearly for any $i, j \in \{0, 1, 2, 3, 4\}$ has the automorphisms $(c, d)\alpha = (c + i, d + j)$ for all $(c, d) \in P$. To see that this is a $PBD(25)$, check to see that each of the differences $(1,0), (2,0), (0,1), (1,1), (2,1), (3,1), (4,1), (0,2), (1,2), (2,2), (3,2)$ and $(4,2)$ occurs in exactly one of the two base blocks (see Exercise 4.2.4).

v = 28: We can construct a $PBD(28)$ using the same technique as when $v = 16$. The following are two orthogonal quasigroups of order 7 (check this!).

$\circ_1$	1	2	3	4	5	6	7
1	1	2	3	4	5	6	7
2	2	3	4	5	6	7	1
3	3	4	5	6	7	1	2
4	4	5	6	7	1	2	3
5	5	6	7	1	2	3	4
6	6	7	1	2	3	4	5
7	7	1	2	3	4	5	6

$\circ_2$	1	2	3	4	5	6	7
1	1	2	3	4	5	6	7
2	3	4	5	6	7	1	2
3	5	6	7	1	2	3	4
4	7	1	2	3	4	5	6
5	2	3	4	5	6	7	1
6	4	5	6	7	1	2	3
7	6	7	1	2	3	4	5

Define a $PBD(28)$ $(\{1,2,\ldots,7\} \times \{1,2,3,4\}, B)$ by defining

$$\begin{aligned} B &= \{\{(x,1),(y,2),(x \circ_1 y,3),(x \circ_2 y,4)\} \mid 1 \leq x,y \leq 7\} \\ &\quad \cup \{\{(1,z),(2,z),\ldots,(7,z)\} \mid 1 \leq z \leq 4\}. \end{aligned}$$

v = 31: Proceed similarly to the case $v = 22$. Start with a $KTS(21)$ with parallel classes $\pi_1,\ldots,\pi_{10}$. For $1 \leq i \leq 10$, to each of the triples in π_i add a new symbol $21 + i$, and add the block $\{22,23,\ldots,31\}$ to produce a $PBD(31)$ on the symbols $1,2,\ldots,31$ with one block of size 10 and the remaining blocks of size 4.

v = 34: The following $PBD(34)$ which uses a more complicated difference method construction was constructed by A. Brouwer. First we construct a $PBD(27)$ with $P = \{1,2,3,\ldots,9\} \times \{1,2,3\}$ and $B = B_1 \cup B_2 \cup B_3 \cup B_4$, where

$$\begin{aligned} B_1 &= \{\{(i,j),(2+i,1+j),(2+i,2+j),(3+i,2+j)\} | 1 \leq i \leq 9, \\ &\quad 1 \leq j \leq 3\}, \\ B_2 &= \{\{(i,j),(3+i,1+j),(5+i,1+j)\} | 1 \leq i \leq 9, 1 \leq j \leq 3\}, \\ B_3 &= \{\{(i,j),(4+i,1+j),(8+i,1+j)\} | 1 \leq i \leq 9, 1 \leq j \leq 3\}, \\ B_4 &= \{\{(i,j),(3+i,j),(6+i,j)\} | 1 \leq i \leq 3, 1 \leq j \leq 3\}, \end{aligned}$$

reducing all sums in the first and second coordinates of each point modulo 9 and 3 respectively. To see that this does define a $PBD(27)$, see Exercise 4.2.5. B_1, B_2, B_3 and B_4 can be represented pictorially as in Figure 4.5.

It is easy to see (Exercise 4.2.6) that the blocks in B_2 can be partitioned into 3 parallel classes of triples π_1, π_2 and π_3; the blocks in B_3 can be partitioned into 3 parallel classes π_4, $\pi_5$and π_6; and the blocks in B_4 are a parallel class π_7. Now add the new symbol $27+i$ to each of the blocks in π_i for $1 \leq i \leq 7$, add the block $\{28,29,\ldots,34\}$, and use the blocks in B_1 to form a $PBD(34)$ with blocks of size 4 and 7 (this last step is similar to the method used when $v = 22$).

v = 37: Define a cyclic $PBD(37)$ with $P = \{1,2,\ldots,37\}$ and $B = \{\{i, 1+i, 3+i, 24+i\}, \{i, 4+i, 9+i, 15+i\}, \{i, 7+i, 17+i, 25+i\} | 1 \leq i \leq 37\}$, reducing all sums modulo 37 (check that each of the difference $1,2,\ldots,$ $(37-1)/2 = 18$ occurs in exactly one of these blocks!).

v = 40: Define a cyclic $PBD(40)$ with $P = \{1,2,\ldots,40\}$ and $B = \{\{i, 1+i, 4+i, 13+i\}, \{i, 2+i, 7+i, 24+i\}, \{i, 6+i, 14+i, 25+i\} | 1 \leq i \leq 40\} \cup \{\{i, 10+i, 20+i, 30+i\} | 1 \leq i \leq 10\}$ reducing all sums modulo 40.

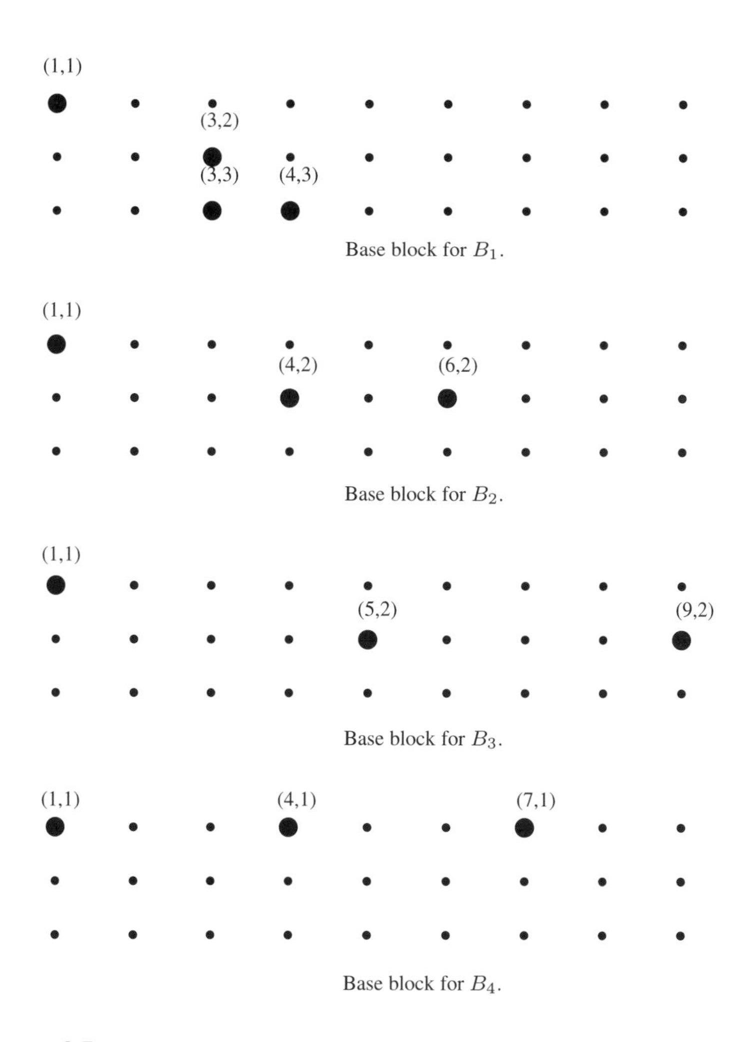

Figure 4.5: *Cycle these 4 base blocks across and down to construct a PBD(27).*

v = 43: For this construction, we introduce a special PBD that is a fundamental "building block" in many constructions in Design Theory. A *group divisible design* is an ordered triple (P, G, B) where P is a finite set, G is a collection of sets called *groups* which partition P, and B is a set of subsets called *blocks* of P, such that $(P, G \cup B)$ is a PBD. The *number* $|P|$ is the *order* of the group divisible design. So a group divisible design is a PBD with a distinguished set of blocks, now called groups, which partition P. If a group divisible design has all groups of the same size, say g, and all blocks of the same size, say k, then we will refer to this design as a $GDD(g, k)$.

For example, in the case $v = 13$ a $PBD(13)$ with all blocks of size 4 was constructed using 2 MOLS(3) (see Figure 4.3). If the point ∞ is deleted from this design then a $GDD(3,4)$ of order 12 results, the groups being $\{(1,\ell),(2,\ell),(3,\ell)\}$ for $1 \leq \ell \leq 4$.

As another example, we can construct a cyclic $GDD(2,4)$ of order 14 by defining

$$\begin{cases} P = & \{1,2,\ldots,14\}, \\ G = & \{\{i,7+i\} | 1 \leq i \leq 7\}, \text{ and} \\ B = & \{\{i,1+i,4+i,6+i\} | 1 \leq i \leq 14\}, \end{cases}$$

reducing all sums modulo 14. (See Figure 4.6).

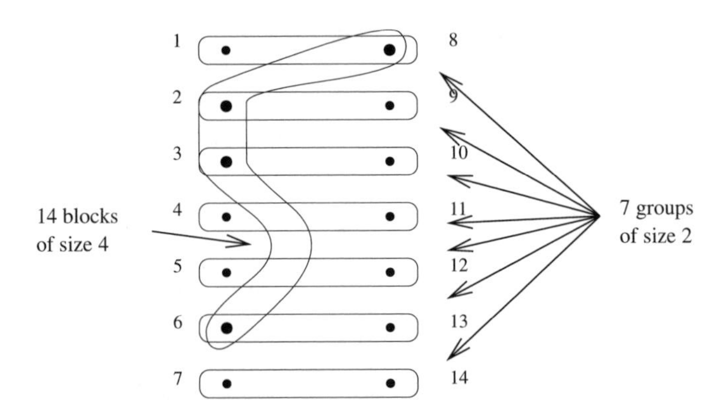

Figure 4.6: *A GDD(2, 4) of order 14.*

We can combine the $GDD(3,4)$ of order 12 and the $GDD(2,4)$ of order 14 constructed above by applying an extremely useful construction: Wilson's Fundamental Construction of GDDs (named after R. M. Wilson). This is so important that we take the time to describe it now.

Wilson's Fundamental Construction. ([26], To construct "big" GDDs from "small" GDDs). Let (P,G,B) be a GDD. Let w be a positive integer called the *weight*, and let $W = \{1,2,\ldots,w\}$. Suppose that for each $b \in B$ there exists a GDD $(W \times b, \{W \times \{p\} \mid p \in b\}, B(b))$. Then $(W \times P, G', B')$ with $G' = \{W \times g \mid g \in G\}$ and $B' = \bigcup_{b \in B} B(b)$, is a GDD.

Notice that if (P,G,B) has group sizes $g_1, g_2, \ldots, g_s$, then $(P \times W, G', B')$ has group sizes $wg_1, wg_2, \ldots, wg_s$, and its block sizes are determined by the sizes of the blocks in $B(b)$, for each $b \in B$.

This construction is exhibited graphically in Figure 4.7.

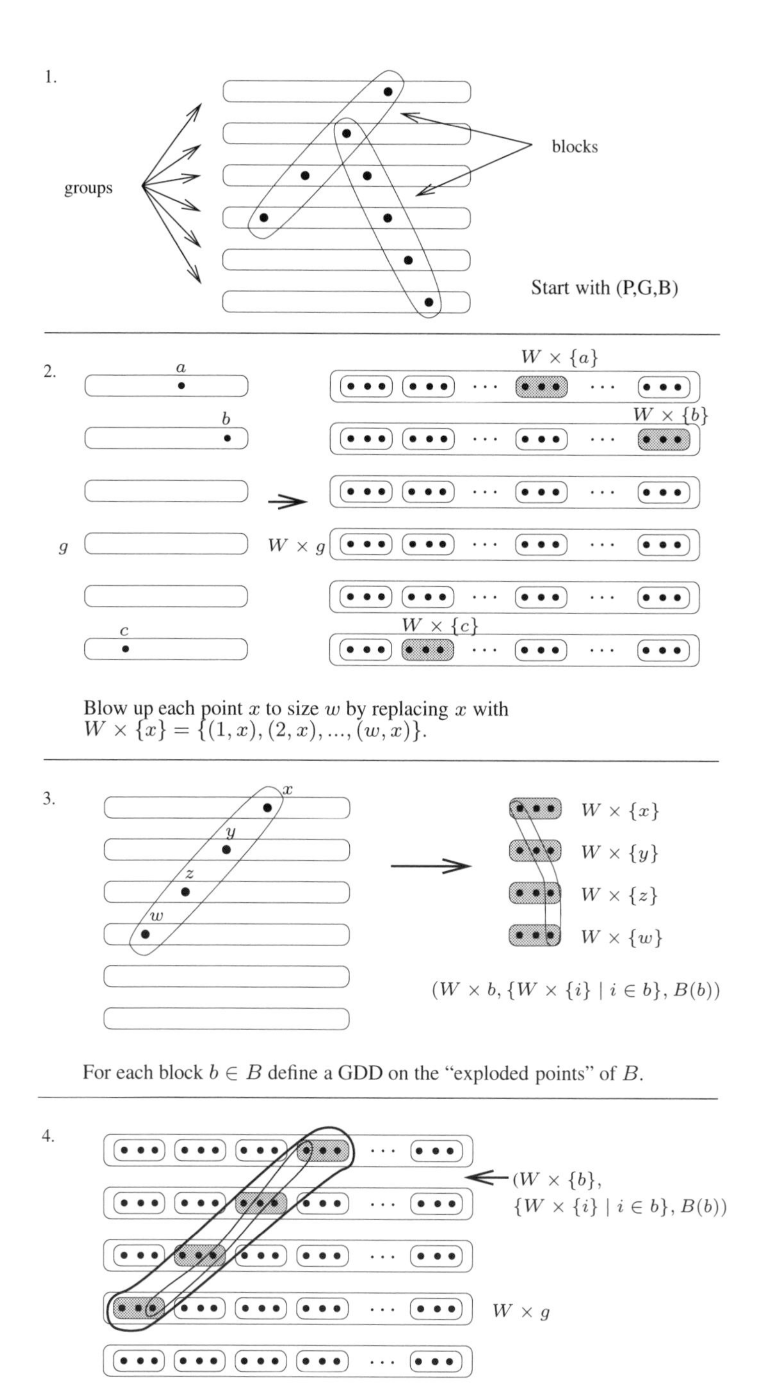

Figure 4.7: *Wilson's Fundamental Construction.*

We now apply Wilson's Fundamental Construction to the $GDD(2,4)$ of order 14 defined above, using a weight of $w = 3$; so $W = \{1,2,3\}$. For each block $b \in B$ we need to define a GDD $(W \times b, \{W \times \{x\} | x \in b\}, B(b))$. Since each block in B has size 4, we can simply use a copy of the $GDD(3,4)$ of order 12 constructed above. Therefore, we end up with a GDD $(\{1,2,3\} \times \{1,\ldots,14\}, \{\{1,2,3\} \times \{i, 7+i\} | 1 \leq i \leq 7\}, \bigcup_{b \in B} B(b))$, which is a $GDD(6,4)$ of order 42. (See Figure 4.8.)

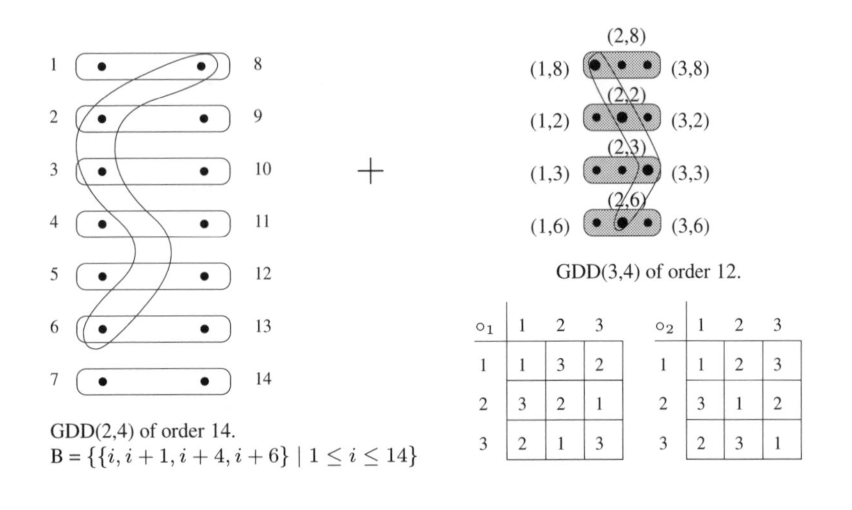

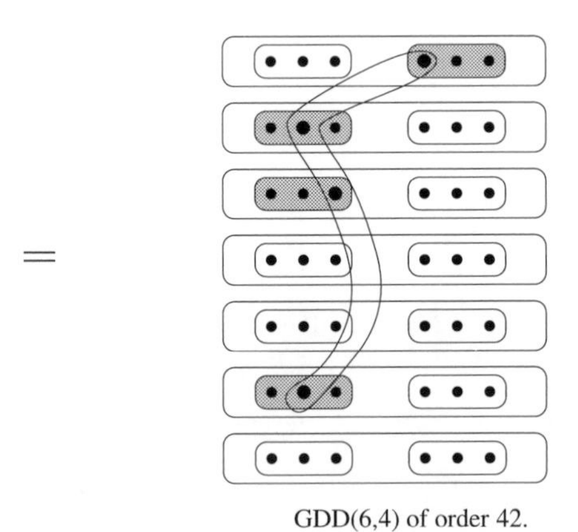

Figure 4.8: *Constructing a* $GDD(6,4)$ *of order* 42.

So, after this big detour, we can finally return to constructing the required $PBD(43)$. We simply add a new point ∞ to each of the groups (of size 6) in the $GDD(6,4)$ of order 42 to produce a $PBD(43)$ in which all blocks have size 4 or 7.

v = 46: Delete any symbol from a $PBD(16)$ in which all blocks have size 4 (see case $v = 16$, for example). The result is a $GDD(3, 4)$ of order 15. Apply Wilson's Fundamental Construction to this GDD using a weight of $w = 3$, and since each block b has size 4, we can define $B(b)$ using the $GDD(3, 4)$ as described in the case $v = 43$. The result is a $GDD(9, 4)$ of order 45. Add a new point ∞ to each of the groups (of size 9) in this $GDD(9, 4)$ to produce a $PBD(46)$ with blocks of size 4 and 10.

v = 79: Define a $GDD(2, 4)$ (P, G, B) of order 26 by letting $P = \{1, 2, \ldots, 26\}, G = \{\{i, 13 + i\} | 1 \leq i \leq 13\}$ and $B = \{\{i, 4 + i, 11 + i, 16+i\}, \{i, 6+i, 8+i, 9+i\} | 1 \leq i \leq 26\}$, reducing all sums modulo 26. Apply Wilson's Fundamental Construction to this GDD using a weight of $w = 3$ and, since each block b has size 4, we can define $B(b)$ using the $GDD(3, 4)$ as described in the case $v = 43$. The result is a $GDD(6, 4)$ of order 78. Add a new point ∞ to each of the groups (of size 6) in this $GDD(6, 4)$ to produce a $PBD(79)$ with blocks of size 4 and 7.

v = 82: In the case $v = 22$, if we delete a point not belonging to the block of size 7 we obtain a $GDD(X, G, B)$, where $|X| = 21$, groups have size 3, and $|b| \in \{4, 7\}$. Now, delete any point from the $PBD(28)$ constructed in the case $v = 28$. The result is a $GDD(P, G^*, B^*)$, where $|P| = 27$, G^* consists of 7 groups of size 3 and 1 group of size 6, and $|b^*| \in \{4, 7\}$. We now apply Wilson's Fundamental Construction to (P, G^*, B^*) using a weight of $w = 3$. For each block of size 7 in B^* we use a copy of (X, G, B) and for each block of size 4 in B^* we use a copy of a $GDD(3, 4)$. The result is a GDD with 7 groups of size 9 and one of size 18, and blocks of size 4 and 7. Adding a new point ∞ to each of the groups gives a $PBD(82)$ with blocks of size $4, 7, 10$, and 19. □

Exercises

4.2.2 The $PBD(22)$ constructed in Lemma 4.2.1 and the recursive construction of Theorem 4.1.3 can be used to find a $KTS(45)$. In this $KTS(45)$, find the triples in the parallel class π_{17}. (To get the answer in the back of the book, for the block $\{v_1, v_2, \ldots, v_7\}$ of size 7 in the PBD(22) with $v_1 < v_2 < \cdots < v_7$, use the KTS(15) in Example 4.1.1, renaming 1 with ∞, $2i$ with $(v_i, 1)$, and $2i + 1$ with $(v_i, 2)$ for $1 \leq i \leq 7$.)

4.2.3 For each of the PBD's constructed in Lemma 4.2.1, count the number of blocks of size $4, 7, 10$ and 19.

4.2.4 For the case $v = 25$ in Lemma 4.2.1, verify that each of the differences $(1, 0), (2, 0), (0, 1), (1, 1), (2, 1), (3, 1), (4, 1), (0, 2), (1, 2), (2, 2),$ $(3, 2)$ and $(4, 2)$ occurs in exactly one of the two base blocks, and deduce that a $PBD(25)$ has indeed been constructed.

4.2.5 For the case $v = 34$ in Lemma 4.2.1, verify that each of the differences $(i, 1)$ for $0 \leq i \leq 9$ and $(0, j)$ for $1 \leq j \leq 4$ occurs in exactly one base block in B, and deduce that a $PBD(27)$ has been constructed.

4.2.6 In the case $v = 34$ in Lemma 4.2.1 partition the triples in B_2 into 3 parallel classes and the triples in B_3 into 3 parallel classes.

We have now finished proving Lemma 4.2.1 and are about ready to use these small PBDs to prove Theorem 4.1.8. The proof we give makes use of the following result. But first a definition. A set of m *mutually orthogonal latin squares* of order n, or m MOLS(n), is a set of m latin squares $\{L_1, L_2, \ldots, L_m\}$ such that for $1 \leq i \neq j \leq m$, L_i is orthogonal to L_j (see Example 5.1.1).

Theorem 4.2.7 *For all $n \geq 4, n \neq 6$ and possibly $n \neq 10$, there exist 3 MOLS(n).* □

Remark In Chapter 5, we prove a similar result, finding the values of n for which there exist 2 MOLS(n). While we never prove Theorem 4.2.7 in this book, the techniques used in Chapter 5 can easily be adapted to obtain such a proof.

For convenience, we restate Theorem 4.1.8 here.

Theorem 4.1.8 For all $n \geq 1$, there exists a $PBD(3n+1)$ with block sizes in $\{4, 7, 10, 19\}$.

Proof The proof is by induction, so we will construct a $PBD(3n + 1)$ with block sizes in $\{4, 7, 10, 19\}$, assuming that for each $s < n$ there exists a $PBD(3s + 1)$ with block sizes in $\{4, 7, 10, 19\}$. In view of Lemma 4.2.1, we can assume that $v = 3n + 1 \geq 49, v \notin \{79, 82\}$. Then it is easy to check that in any case, we can express v as

$$v = 12m + 3t + 1, \text{ where } 0 \leq t \leq m, m \geq 4 \text{ and } m \notin \{6, 10\}$$

(Check this!). By Theorem 4.2.7, there exist 3 MOLS(m), L_1, L_2 and L_3 with corresponding quasigroups $(S, \circ_1)$, $(S, \circ_2)$, and $(S, \circ_3)$, where $S = \{1, 2, \ldots, m\}$. Following the type of construction in the case $v = 16$ of Lemma 4.2.1, we define a $GDD(m, 5)$ (P', G', B') of order $5m$ with

$$\begin{aligned} P' &= S \times \{1, 2, 3, 4, 5\}, \\ G' &= \{S \times \{i\} | 1 \leq i \leq 5\}, \text{ and} \\ B' &= \{\{(x, 1), (y, 2), (x \circ_1 y, 3), (x \circ_2 y, 4), (x \circ_3 y, 5)\} | x, y \in S\}. \end{aligned}$$

Modify this $GDD(m, 5)$ by deleting exactly $m - t$ points, all having the same second coordinate, say 5 (note that this is possible since $0 \leq t \leq m$). This leaves a GDD (P, G, B), with $|P| = 4m + t$, 4 groups of size m and one group of size t, and blocks of size 5 and 4 (the blocks of size 4 coming from blocks of size 5 that contained a deleted point). To this GDD we apply Wilson's Fundamental Construction using a weight of $w = 3$. For each block $b \in B$ of size 4, we define $B(b)$ using the $GDD(3, 4)$ of order 12 obtained from 2 MOLS(3) (see case $v = 43$ of Lemma 4.2.1). For each block $b \in B$ of size 5, we define $B(b)$ using the $GDD(3, 4)$ of order 15 formed by deleting one point from a $PBD(16)$ in which all blocks have size 4 (see case $v = 16$ of Lemma 4.2.1).

Since $B(b)$ contains only blocks of size 4, for all $b \in B$, Wilson's Fundamental Construction produces a GDD of order $3|P| = 12m+3t$ with all blocks of size 4, 4 groups of size $3m$, and one group of size $3t$.

Finally we add a new point ∞, which is added to each group, producing 4 blocks of size $3m+1$ and one block of size $3t+1$. By induction, we can replace each of these blocks with a PBD with block sizes in $4, 7, 10$ and 19, and so the result follows. □

Example 4.2.8 We construct a $PBD(178)$ with block sizes in $\{4, 7, 10, 19\}$ using the proof of Theorem 4.1.8 just completed, and the small PBDs constructed in Lemma 4.2.1. Figure 4.9 should help the reader.

We begin by suitably choosing $m = 14$ and $t = 3$ (so $178 = 12m + 3t + 1, 0 \leq t \leq m, m \geq 4$ and $m \notin \{6, 10\}$). We construct the GDD (P', G', B') using 3 MOLS(14) and then delete $m-t = 11$ points, all with second coordinate 5. Applying Wilson's Fundamental Theorem with weight $w = 3$ gives a GDD with 4 groups of size $3m = 42$ and one group of size $3t = 9$. Adding ∞ to each of these groups, then using the $PBD(3m+1 = 43)$ and the $PBD(3t+1 = 10)$ constructed in Lemma 4.2.1, produces the required $PBD(178)$. Notice that the $PBD(43)$ has 7 blocks of size 7 and the $PBD(10)$ has one block of size 10, so the $PBD(178)$ has 28 blocks of size 7, one block of size 10, and the rest of the blocks have size 4. For a pictorial representation of this construction see Figure 4.9.

Exercises

4.2.9 For the following values of v, a $PBD(v)$ with block sizes in $\{4, 7, 10, 19\}$ is constructed using the proof of Theorem 4.1.8 and the small PBDs constructed in Lemma 4.2.1. Find suitable values of m and t, then count the number of blocks of each size in the $PBD(v)$. (The number of blocks of each size for the $PBD(3n+1)$s with $1 \leq n \leq 15$ and with $26 \leq n \leq 27$ were counted in Exercise 4.2.3.)

(a) 97
(b) 103
(c) 142
(d) 190

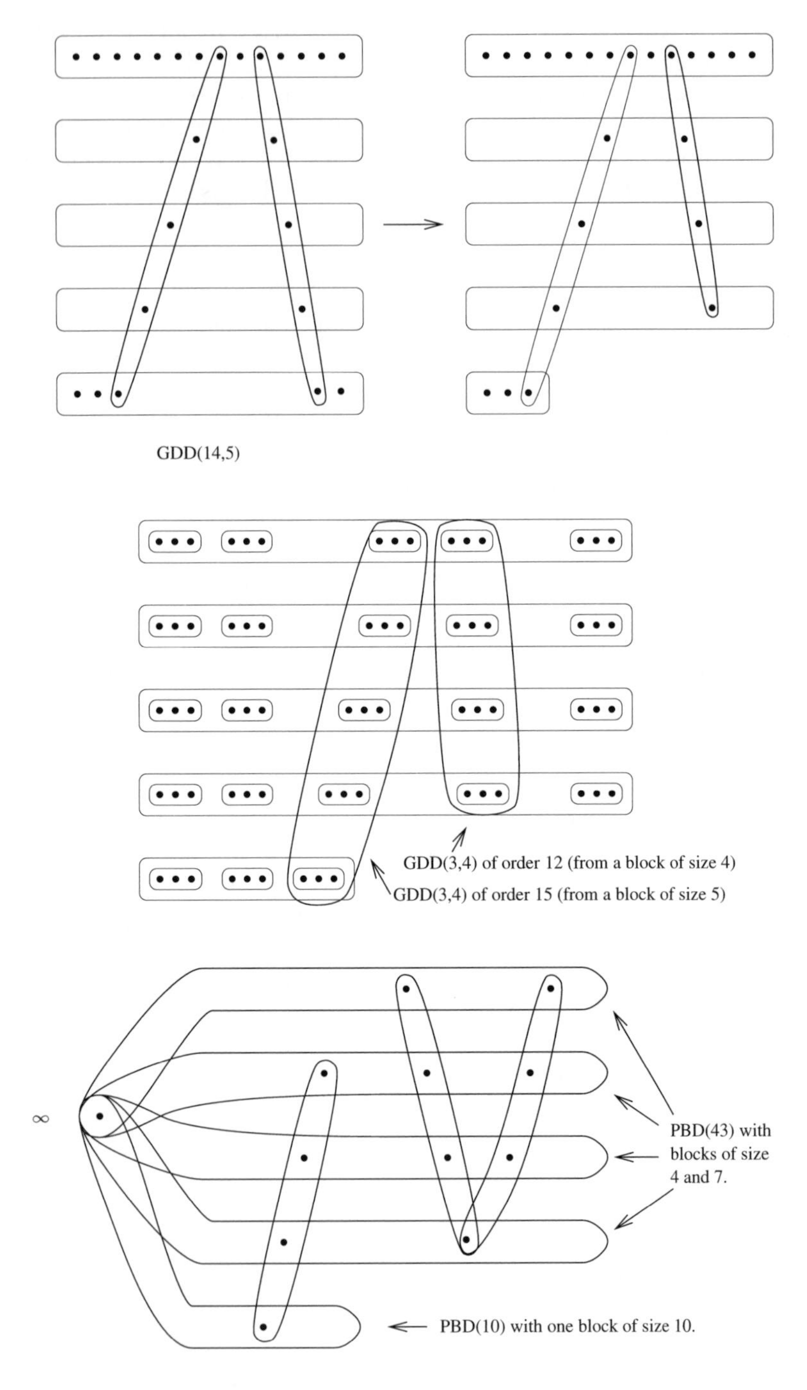

Figure 4.9: *Constructing a* $PBD(178)$.

Chapter 5

Mutually Orthogonal Latin Squares

5.1 Introduction

Two latin squares L_1 and L_2 of order n are said to be *orthogonal* if for each $(x, y) \in \{1, 2, \ldots, n\} \times \{1, 2, \ldots, n\}$ there is exactly one ordered pair (i, j) such that cell (i, j) of L_1 contains the symbol x and cell (i, j) of L_2 contains the symbol y. In other words, if L_1 and L_2 are superimposed, the resulting set of n^2 ordered pairs are *distinct*. The latin squares $L_1, L_2, \ldots, L_t$ are said to be *mutually orthogonal* if for $1 \leq a \neq b \leq t$, L_a and L_b are orthogonal.

The easiest way to show that a pair of latin squares are orthogonal is to use the "famous" *two finger rule*.

The Two Finger Rule. Let L and M be latin squares of order n. Then L and M are orthogonal if and only if whenever a pair of cells are occupied by the *same* symbol in L, they are occupied by *different* symbols in M.

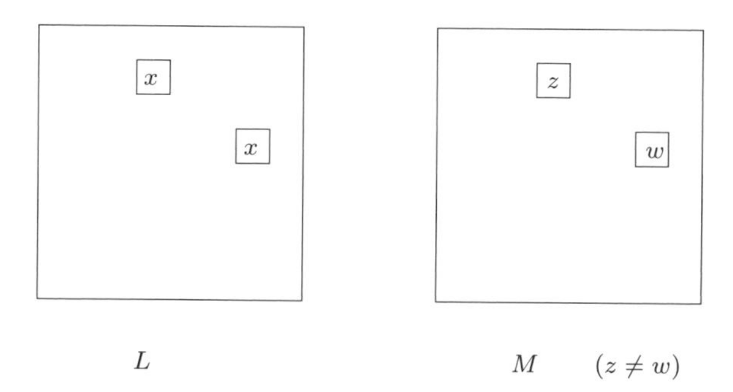

Figure 5.1: *The Two Finger Rule.*

Example 5.1.1 a) L_1 and L_2 are orthogonal latin squares of order 3, where

$L_1 =$

1	3	2
3	2	1
2	1	3

and $L_2 =$

1	2	3
3	1	2
2	3	1

.

b) L_3, L_4 and L_5 are three mutually orthogonal latin squares of order 4, where

$L_3 =$

1	3	4	2
4	2	1	3
2	4	3	1
3	1	2	4

$L_4 =$

1	4	2	3
3	2	4	1
4	1	3	2
2	3	1	4

and $L_5 =$

1	2	3	4
2	1	4	3
3	4	1	2
4	3	2	1

.

The study of pairs of orthogonal latin squares goes back to 1782 when Euler [7] considered the following problem.

The Euler Officer Problem Six officers from each of six different regiments are selected so that the six officers from each regiment are of six different ranks, the same six ranks being represented by each regiment. Is it possible to arrange these 36 officers in a 6×6 array so that each regiment and each rank is represented exactly once in each row and column of this array?

If we number the ranks 1, 2, 3, 4, 5 and 6, and number the regiments 1, 2, 3, 4, 5 and 6, then each officer is represented by a unique ordered pair $(x, y) \in \{1, 2, 3, 4, 5, 6\} \times \{1, 2, 3, 4, 5, 6\}$, the first coordinate being his rank, the second his regiment. Using this representation it is clear that Euler's Officer Problem asks if it is possible to arrange the 36 ordered pairs (x, y); $x, y \in \{1, 2, 3, 4, 5, 6\}$; on a 6×6 grid so that in each row and column the first coordinates are $1, 2, 3, 4, 5$

and 6 in some order *and* the second coordinates are $1, 2, 3, 4, 5, 6$ in some order.

11	12	13	14	15	16
21	22	23	24	25	26
31	32	33	34	35	36
41	42	43	44	45	46
51	52	53	54	55	56
61	62	63	64	65	66

$\longrightarrow$

Figure 5.2: *The Euler Officer Problem: can the 36 ordered pairs formed from* $1, 2, 3, 4, 5, 6$ *be arranged on a* 6×6 *grid so that the first coordinates are different in* each *row and column and the second coordinates are different in* each *row and column.*

Clearly, if a solution to the Euler Officer Problem exists, the first coordinates form a latin square *and* the second coordinates form a latin square. (Figure 5.3). Hence the Euler Officer Problem is equivalent to a pair of orthogonal latin squares of order 6.

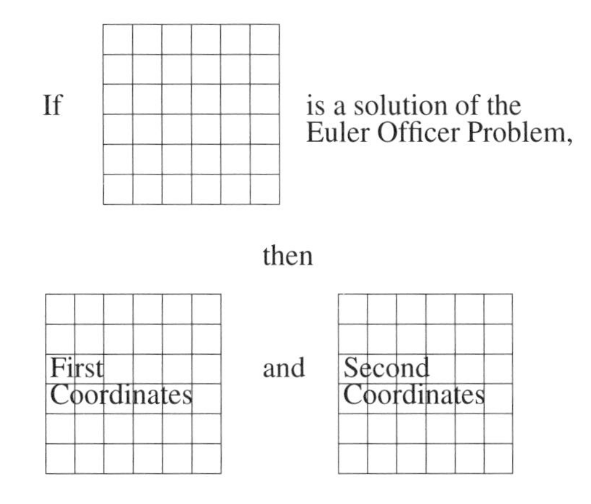

Figure 5.3: *Solution of the Euler Officer Problem.*

However in 1900 G. Tarry [24] used brute force to show that no such pair of orthogonal latin squares exists! For a non-brute force proof see Doug Stinson [23].

A latin square of order n is in *standard form* if for $1 \leq i \leq n$ cell $(1, i)$ contains the symbol i. If L_1 and L_2 are orthogonal, and if L_1' and L_2' are produced from L_1 and L_2 respectively by renaming the symbols, then clearly L_1' and L_2' are also orthogonal. (See Exercise 5.1.2.) Therefore, if we have a set of mutually orthogonal latin squares of order n (MOLS(n)), we can always rename the symbols in each latin square so that each of the squares in the resulting set of MOLS(n) is in standard form.

If $\{L_1, \ldots, L_t\}$ is a set of MOLS(n), it is natural to ask how large t can be. Since we can assume that each of $L_1, \ldots, L_t$ is in standard form, we can quickly obtain an upper bound on t by considering the symbols in cell $(2, 1)$ of $L_1, \ldots, L_t$ (so we are assuming that $n \geq 2$).

Now, the symbol 1 occurs in cell $(1, 1)$ of each of these latin squares, so the cell $(2, 1)$ in each of them must be occupied by a symbol from the set $\{2, 3, 4, \ldots, n\}$. Furthermore, when any two of these latin squares are superimposed, since they are in standard form, the first rows give the ordered pairs $(1, 1), (2, 2), (3, 3), (4, 4), \ldots, (n, n)$. Hence the cell $(2, 1)$ *cannot* be occupied by the same symbol x in different squares since the ordered pair (x, x) would occur twice when they are superimposed (see Figure 5.4). So the symbols in cell $(2, 1)$ of $L_1, \ldots, L_t$ are all different and belong to the set $\{2, 3, \ldots, n\}$. Therefore, if $n \geq 2$ then $t \leq n - 1$. Clearly this upper bound can be obtained, since 2 MOLS(3) and 3 MOLS(4) are exhibited in Example 5.1.1. A *complete set* of MOLS(n) is defined to be a set of $n - 1$ MOLS(n). In the next section we will give a construction for complete sets of MOLS(n) when n is a prime power.

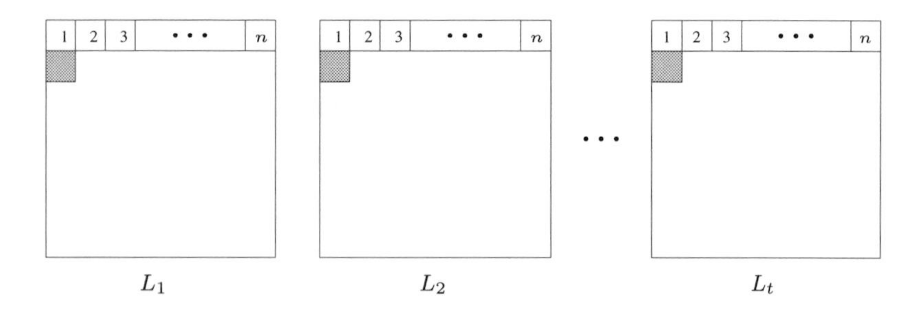

Figure 5.4: *MOLS(n) in standard form. The cell (2, 1) must be occupied by a symbol from the set $\{2, 3, \ldots, n\}$ and no two squares can have the same symbol in cell (2, 1).*

Exercises

5.1.2 If L is a latin square of order n and α is any permutation on

$\{1, 2, 3, \ldots, n\}$, denote by $L\alpha$ the latin square obtained from L by replacing each symbol x in L by $x\alpha$. Prove that if L_1 and L_2 are orthogonal *and* α and β are any two permutations, then $L_1\alpha$ and $L_2\beta$ are also orthogonal.

5.1.3 Apply a permutation to the symbols in L_3, L_4 and L_5 of Example 5.1.1 to produce a complete set of MOLS(4) in standard form.

5.1.4 Show that if $\{L_1, \ldots, L_t\}$ is a set of MOLS(n), all of which are idempotent, then $t \leq n - 2$. (Notice that 2 of the MOLS(4) in Example 5.1.1 are idempotent, so this bound can be attained.)

5.1.5 Show that any set of $n-2$ MOLS(n) can be extended to a complete set of MOLS(n).

5.2 The Euler and MacNeish Conjectures

As mentioned in Section 5.1, in 1782 Euler considered the problem of finding two orthogonal latin squares of order 6. At that time he made the following conjecture.

The Euler Conjecture. A pair of orthogonal latin squares of order n exists if and only if $n \equiv 0, 1,$ or $3 \pmod 4$.

Euler made this conjecture because he was able to construct a pair of orthogonal latin squares of every order $n \equiv 0, 1,$ or $3 \pmod 4$ and was *not* able to construct such a pair for any $n \equiv 2 \pmod 4$.

In 1900 G. Tarry [24] proved that there did not exist a pair of orthogonal latin squares of order 6. (Tarry used brute force. For a non brute force proof see Doug Stinson [23].)

In 1922 H. MacNeish [16] made the following conjecture.

The MacNeish Conjecture. Let $n = p_1^{r_1} p_2^{r_2} \cdots p_x^{r_x}$, where each of $p_1, p_2, p_3, \cdots, p_x$ is a distinct prime. Then the maximum number of MOLS(n) is $m(n) = \min\{p_1^{r_1}, p_2^{r_2}, p_3^{r_3}, \ldots, p_x^{r_x}\} - 1$.

Clearly the Euler Conjecture is a special case of the MacNeish Conjecture, since if $n \equiv 2 \pmod 4$, $n = 2\cdot(\text{odd number}) = 2\cdot p_2^{r_2} p_3^{r_3} \cdots p_x^{r_x}$, and so $m(n) = \min\{2, p_2^{r_2} p_3^{r_3}, \ldots, p_x^{r_x}\} - 1 = 2 - 1 = 1$.

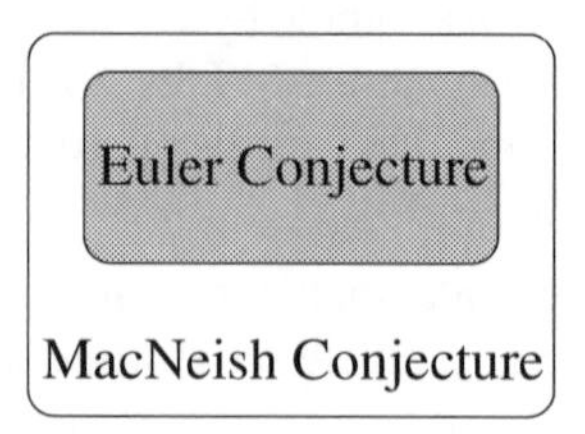

Figure 5.5: *The Euler Conjecture is a special case of the MacNeish Conjecture.*

MacNeish had a very good reason for his conjecture. He found a *finite field construction* and a *direct product construction* which enabled him to construct $m(n)$ MOLS(n) for every n. (Finite fields had not been invented in Euler's day.)

The first inkling that the Euler Conjecture might not be true came in 1957 when E. T. Parker constructed 3 MOLS(21), thereby killing the MacNeish Conjecture. Since the Euler Conjecture is a special case of the MacNeish Conjecture it was not particularly surprising when, in 1958, R. C. Bose and S. S. Shrikhande [2] constructed a pair of orthogonal latin squares of order 22, and in so doing disproved the long standing Euler Conjecture. Shortly after this "event", E. T. Parker [17] constructed a pair of orthogonal latin squares of order 10. Finally, in 1960 R. C. Bose, S. S. Shrikhande, and E. T. Parker [3] proved that a pair of orthogonal latin squares of order n exists for all n other than 2 or 6 (for which no such pair exists).

In what follows, we will give a systematic verification of all of the above existence results. It will be convenient to use quasigroup notation to define orthogonal latin squares in much of what follows.

Everything that follows in predicated on a *reasonable knowledge* of finite fields. Among other things, how to construct them, as well as the fact that the *spectrum* (= the set of all *orders* for which a finite field exists) is precisely the set of all p^r, where p is a prime and r is a positive integer.

The Finite Field Construction. Let $(F, +, \cdot)$ be a finite field of order n. Since there is nothing sacred about the symbols representing the field elements in F, it is convenient to assume that $F = \{1, 2, 3, \ldots, n\}$ with the proviso the n represents the zero field element. For each $n \neq k \in F$, define a binary operation $\circ(k)$ on F by $x \circ(k) y = x \cdot k + y$ (arithmetic computed in the finite field $(F, +, \cdot)$. Then $(F, \circ(k))$ is a *quasigroup* (latin square).

It is easy to see that $(F, \circ(k))$ is in fact a quasigroup. Suppose $x \circ (k) y = x \circ (k) z$. Then $x \cdot k + y = x \cdot k + z$ and so $y = z$. On the other hand if $y \circ (k) x = z \circ (k) x$, then $y \cdot k + x = z \cdot k + x$, so $y \cdot k = z \cdot k$, and since $k \neq$ zero we get $y = z$.

Example 5.2.1 A quasigroup of order 4 constructed using the Finite Field Construction. We first construct the finite field of order 4 using the irreducible polynomial $1 + x + x^2$.

+	0	1	x	1+x
0	0	1	x	1+x
1	1	0	1+x	x
x	x	1+x	0	1
1+x	1+x	x	1	0

·	0	1	x	1+x
0	0	0	0	0
1	0	1	x	1+x
x	0	x	1+x	1
1+x	0	1+x	1	x

Renaming the field elements $0, 1, x$, and $1+x$ with $4, 1, 2$, and 3 respectively gives the tables

+	4	1	2	3
4	4	1	2	3
1	1	4	3	2
2	2	3	4	1
3	3	2	1	4

·	4	1	2	3
4	4	4	4	4
1	4	1	2	3
2	4	2	3	1
3	4	3	1	2

Since there are 3 nonzero field elements $1, 2$, and 3, we can construct 3 quasigroups as follows:

∘(1)	4	1	2	3
4	4	1	2	3
1	1	4	3	2
2	2	3	4	1
3	3	2	1	4

$x \cdot 1 + y$

∘(2)	4	1	2	3
4	4	1	2	3
1	2	3	4	1
2	3	2	1	4
3	1	4	3	2

$x \cdot 2 + y$

$\circ(3)$	4	1	2	3
4	4	1	2	3
1	3	2	1	4
2	1	4	3	2
3	2	3	4	1

$x \cdot 3 + y$

(Using the addition and multiplication tables of the finite field of order 4 given above).

Theorem 5.2.2 *Let $(F, +, \cdot)$ be a finite field of order n, where $F = \{1, 2, 3, \ldots, n = \text{ zero}\}$. Then the quasigroups $(F, \circ(1))$, $(F, \circ(2)), \ldots, (F, \circ(n-1))$ are a complete set of mutually orthogonal quasigroups (latin squares) of order n.*

Proof It is convenient to use latin square vernacular in the proof. We denote the latin square associated with the quasigroup $(F, \circ(k))$ by $L(k)$. We will check that $L(k)$ and $L(\ell)$ are orthogonal for $1 \leq k < \ell < n$ by using The Two Finger Rule.

Suppose that cells (i_1, j_1) and (i_2, j_2) both contain the same symbol in $L(k)$ and both contain the same symbol in $L(\ell)$. We will prove that $i_1 = i_2$ and $j_1 = j_2$. For convenience, let $x \cdot y$ be denoted by xy.

By definition of $(F, \circ(k))$ and $(F, \circ(\ell))$ we have that

$$\begin{aligned} i_1 k + j_1 &= i_2 k + j_2, \text{ and} \\ i_1 \ell + j_1 &= i_2 \ell + j_2. \end{aligned}$$

So

$$i_1 k + j_1 - i_2 k = j_2 = i_1 \ell + j_1 - i_2 \ell, \text{ or}$$

$$(i_1 - i_2)k = (i_1 - i_2)\ell.$$

Since $k \neq \ell$ and $k, \ell \neq$ zero, this means that $i_1 - i_2 =$ zero, so $i_1 = i_2$ and therefore $j_1 = j_2$. □

Example 5.2.3 The 3 quasigroups in Example 5.2.1 constructed using The Finite Field Construction are mutually orthogonal (check it out!).

Example 5.2.4 We construct 4 MOLS(5) using the finite field construction. The finite field of order 5 is, of course, the integers (mod 5). So we can easily get the following 4 orthogonal quasigroups of order 5 by defining $i \circ (k) j = ik + j$ (mod 5), renaming zero with 5.

$\circ(1)$	5	1	2	3	4
5	5	1	2	3	4
1	1	2	3	4	5
2	2	3	4	5	1
3	3	4	5	1	2
4	4	5	1	2	3

$i \cdot 1 + j$

$\circ(2)$	5	1	2	3	4
5	5	1	2	3	4
1	2	3	4	5	1
2	4	5	1	2	3
3	1	2	3	4	5
4	3	4	5	1	2

$i \cdot 2 + j$

$\circ(3)$	0	1	2	3	4
5	5	1	2	3	4
1	3	4	5	1	2
2	1	2	3	4	5
3	4	5	1	2	3
4	2	3	4	5	1

$i \cdot 3 + j$

$\circ(4)$	5	1	2	3	4
5	5	1	2	3	4
1	4	5	1	2	3
2	3	4	5	1	2
3	2	3	4	5	1
4	1	2	3	4	5

$i \cdot 4 + j$

Example 5.2.5 The finite field $(F, +, \cdot)$ of order 25 can be constructed using the irreducible *primitive* polynomial $p(x) = 3 + 2x + x^2$. To facilitate the arithmetic in $(F, +, \cdot)$, we represent each non-zero field element as a power of the primitive element x as follows:

i	x^i	i	x^i	i	x^i	i	x^i
0	1	6	3	12	4	18	2
1	x	7	3x	13	4x	19	2x
2	2 + 3x	8	1 + 4x	14	3 + 2x	20	4 + x
3	1 + x	9	3 + 3x	15	4 + 4x	21	2 + 2x
4	2 + 4x	10	1 + 2x	16	3 + x	22	4 + 3x
5	3 + 4x	11	4 + 2x	17	2 + x	23	1 + 3x

a) Find the cell in "column" x^9 of $(F, \circ(x^{19}))$ that contains the symbol x^{11}.

By definition, cell (i, j) of $(F, \circ(k))$ contains symbol $ik + j$. We need to find the row i, and we know that

$$\begin{aligned} ix^{19} + x^9 &= x^{11}, \text{ so} \\ ix^{19} &= x^{11} - x^9 \\ &= (4 + 2x) - (3 + 3x) \\ &= 1 + 4x, \text{ so} \\ i &= x^8 / x^{19} \\ &= x^{13} \end{aligned}$$

(since $x^{24} = 1$, so $x^{-11} = x^{-11}x^{24} = x^{13}$). Therefore cell (x^{13}, x^9) of $(F, \circ(x^{19}))$ contains symbol x^{11}.

b) Find the cell that contains the symbol x^{20} in $(F, \circ(x^6))$ and the symbol x^{10} in $(F, \circ(x^{14}))$.

Now we have to find i and j, where

$$\begin{aligned} ix^6 + j &= x^{20}, \text{ and} \\ ix^{14} + j &= x^{10}. \end{aligned}$$

Solving these simultaneously, we get

$$\begin{aligned} i(x^{14} - x^6) &= x^{10} - x^{20}, \text{ so} \\ i &= ((1 + 2x) - (4 + x))/((3 + 2x) - (3)) \\ &= (2 + x)/2x \\ &= x^{17}/x^{19} \\ &= x^{22}, \end{aligned}$$

(again, since $x^{24} = 1$). Substituting for i gives

$$\begin{aligned} x^{22}x^6 + j &= x^{20}, \text{ so} \\ j &= x^{20} - x^{28} \\ &= (4 + x) - (2 + 4x) \\ &= x^{21}. \end{aligned}$$

Therefore we have that cell (x^{22}, x^{21}) contains the symbol x^{20} in $(F, \circ(x^6))$ and contains the symbol x^{10} in $(F, \circ(x^{14}))$.

Exercises

5.2.6 Form addition and multiplication tables for the finite field of order 7. Use these to construct 6 MOLS(7).

5.2.7 Form addition and multiplication tables for the finite field of order 8, constructing the field with the irreducible polynomial $p(x) = 1 + x + x^3$. Rename the field elements with the symbols $1, \ldots, 8$ and construct 7 MOLS(8).

5.2.8 Form addition and multiplication tables for the finite field of order 9, constructing the field with the irreducible polynomial $p(x) = 2 + 2x + x^2$. Rename the field elements with the symbols $1, \ldots, 9$ and construct 8 MOLS(9).

5.2.9 Continuing Example 5.2.5(a), find the cell in

(a) row x^3 of $L(x^{12})$ that contains symbol x^{15},
(b) row x^4 of $L(x^{16})$ that contains symbol $x^0 = 1$,
(c) column 0 of $L(x^9)$ that contains symbol x^7, and
(d) column x^{19} of $L(x^{21})$ that contains symbol x^{23}.

5.2.10 Continuing Example 5.2.5(b), find the cell that contains

(a) symbol x^4 in $L(x^4)$ and symbol x^8 in $L(x^{10})$,
(b) symbol x^0 in $L(x^2)$ and symbol x^{14} in $L(x^{20})$, and
(c) symbol x^3 in $L(x^3)$ and symbol 0 in $L(x^{18})$.

5.2.11 Let L be a latin square of order n and let r be any permutation on $\{1, 2, 3, \ldots, n\}$. Denote by rL the latin square obtained from L by permuting the rows of L according to r. Let $L_1, L_2, \ldots, L_t$ be t MOLS(n) and r any permutation of the rows of L_1 so that rL_1 has a *constant* main diagonal. Then the latin squares $rL_1, rL_2, \ldots, rL_t$ are t MOLS(n) and furthermore each of $rL_2, rL_3, \ldots, rL_t$ is *diagonalized* (= no 2 symbols on the main diagonal are the same). Applying appropriate permutations $\alpha_2, \alpha_3, \ldots, \alpha_t$ to the symbols in each of $rL_2, rL_3, \ldots, rL_t$ gives $t - 1$ *idempotent* MOLS(n): $rL_2\alpha_2, rL_3\alpha_3, \ldots, rL_t\alpha_t$. Use the above algorithm to construct 2 idempotent MOLS(4), 3 idempotent MOLS(5), and 5 idempotent MOLS(7). (See Examples 5.2.1 and 5.2.4 and Exercise 5.2.6 respectively.)

5.2.12 Prove that there exists $m(n) - 1$ idempotent MOLS(n).

We now present the direct product construction.

The Direct Product Construction. The *direct product* $A \times B$ of two latin squares A and B of orders m and n respectively is the latin square of order mn defined by

$A \times B$ =

A(1,1)	A(1,2)	• • •	A(1,n)
A(2,1)	A(2,2)	• • •	A(2,n)
⋮	⋮	⋱	⋮
A(n,1)	A(n,2)	• • •	A(n,n)

where $A(i, j)$ is the latin square of order m formed from A by replacing each entry x with (x, k), where k is the entry in cell (i, j) of B.

Example 5.2.13 The direct product of *latin squares* of orders 3 and 4.

$A =$

3	1	2
2	3	1
1	2	3

$B =$

1	3	4	2
4	2	1	3
2	4	3	1
3	1	2	4

$A \times B =$

3 1	1 1	2 1	3 3	1 3	2 3	3 4	1 4	2 4	3 2	1 2	2 2
2 1	3 1	1 1	2 3	3 3	1 3	2 4	3 4	1 4	2 2	3 2	1 2
1 1	2 1	3 1	1 3	2 3	3 3	1 4	2 4	3 4	1 2	2 2	3 2
3 4	1 4	2 4	3 2	1 2	2 2	3 1	1 1	2 1	3 3	1 3	2 3
2 4	3 4	1 4	2 2	3 2	1 2	2 1	3 1	1 1	2 3	3 3	1 3
1 4	2 4	3 4	1 2	2 2	3 2	1 1	2 1	3 1	1 3	2 3	3 3
3 2	1 2	2 2	3 4	1 4	2 4	3 3	1 3	2 3	3 1	1 1	2 1
2 2	3 2	1 2	2 4	3 4	1 4	2 3	3 3	1 3	2 1	3 1	1 1
1 2	2 2	3 2	1 4	2 4	3 4	1 3	2 3	3 3	1 1	2 1	3 1
3 3	1 3	2 3	3 1	1 1	2 1	3 2	1 2	2 2	3 4	1 4	2 4
2 3	3 3	1 3	2 1	3 1	1 1	2 2	3 2	1 2	2 4	3 4	1 4
1 3	2 3	3 3	1 1	2 1	3 1	1 2	2 2	3 2	1 4	2 4	3 4

The direct product of the quasigroups $(A, \circ_1)$ and $(B, \circ_2)$ is just the direct product of the latin squares A and B *with* appropriate headlines and sidelines. Algebraically, the direct product is defined to be the quasigroup $(A \times B, \odot)$ where $(x, y) \odot (z, w) = (x \circ_1 z, y \circ_2 w)$.

Example 5.2.14 The direct product of the *quasigroups* corresponding to the latin squares A and B in Example 5.2.13.

$(A, \circ_1) =$

$\circ_1$	1	2	3
1	3	1	2
2	2	3	1
3	1	2	3

$(B, \circ_2) =$

$\circ_2$	1	2	3	4
1	1	3	4	2
2	4	2	1	3
3	2	4	3	1
4	3	1	2	4

$(A \times B, \odot) =$

$\odot$	11	21	31	12	22	32	13	23	33	14	24	34
11	31	11	21	33	13	23	34	14	24	32	12	22
21	21	31	11	23	33	13	24	34	14	22	32	12
31	11	21	31	13	23	33	14	24	34	12	22	32
12	34	14	24	32	12	22	31	11	21	33	13	23
22	24	34	14	22	32	12	21	31	11	23	33	13
32	14	24	34	12	22	32	11	21	31	13	23	33
13	32	12	22	34	14	24	33	13	23	31	11	21
23	22	32	12	24	34	14	23	33	13	21	31	11
33	12	22	32	14	24	34	13	23	33	11	21	31
14	33	13	23	31	11	21	32	12	22	34	14	24
24	23	33	13	21	31	11	22	32	12	24	34	14
34	13	23	33	11	21	31	12	22	32	14	24	34

Theorem 5.2.15 *If A_1 and A_2 are orthogonal latin squares of order m and B_1 and B_2 are orthogonal latin squares of order n, then the direct products $A_1 \times B_1$ and $A_2 \times B_2$ are orthogonal latin squares of order mn.*

Proof In what follows we will index the rows and columns with the mn ordered pairs $(x, y) \in \{1, 2, 3, \ldots, m\} \times \{1, 2, \ldots, n\}$. We will use the Two Finger Rule. So, suppose cells $((x_1, y_1), (z_1, w_1))$ and $((x_2, y_2), (z_1, w_2))$ contain the same symbol (a, b) in $A_1 \times B_1$. We will show that these cells are occupied by different symbols in $A_2 \times B_2$. There are two cases to consider: (1) $(y_1, w_1) = (y_2, w_2)$ and (2) $(y_1, w_1) \neq (y_2, w_2)$. We handle each in turn.

(1) $(y_1, w_1) = (y_2, w_2)$: In this case the entry in cells (y_1, w_1) of *both* B_1 and B_2 is "b". Now the first coordinates of the entries in cells $((x_1, y_1), (z_1, w_1))$ and $((x_2, y_2), (z_2, w_2))$ are determined in A_1 and A_2 by the entries in cells (x_1, z_1) and (x_2, z_2). Since these cells are occupied by "a" in A_1, they must be occupied by *distinct* symbols in A_2, say "c" and "d", and so the cells $((x_1, y_1), (z_1, w_1))$ and $((x_2, y_2), (z_2, w_2))$ in $A_2 \times B_2$ are occupied by (c, e) and (d, e), where $c \neq d$. See Figure 5.6.

(2) $(y_1, w_1) \neq (y_2, w_2)$: In this case, since cells (y_1, w_1) and (y_2, w_2) are occupied by b in B_1 and they are *not* the same cell, we must have $y_1 \neq y_2$ *and* $w_1 \neq w_2$. But since B_1 and B_2 are orthogonal, these cells must be occupied by distinct symbols, say "e" and "f", in B_2. Hence the cells $((x_1, y_1), (z_1, w_1))$

and $((x_2, y_2), (z_2, w_2))$ in $A_2 \times B_2$ are occupied by (u, e) and (w, f), where $e \neq f$ ((u and w may or may not be equal and cells (x_1, z_1) and (x_2, z_2) may or may not be the same; it doesn't matter). See Figure 5.7.

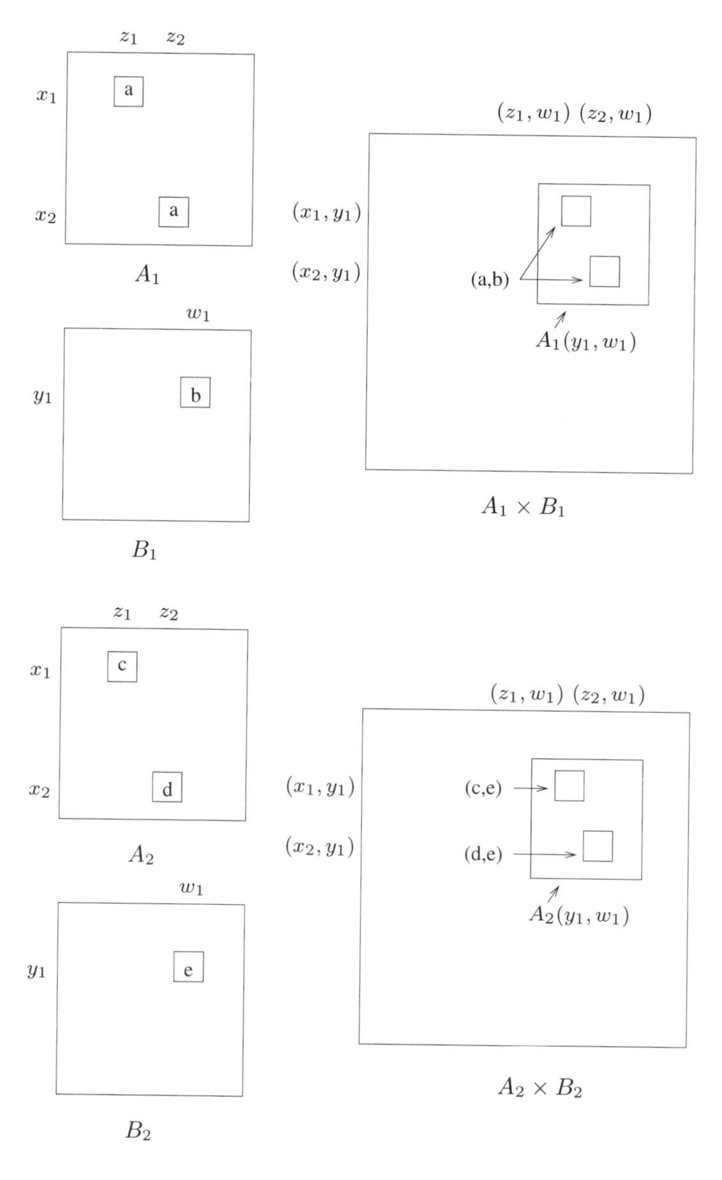

Figure 5.6: *The direct product of MOLS: case 1.*

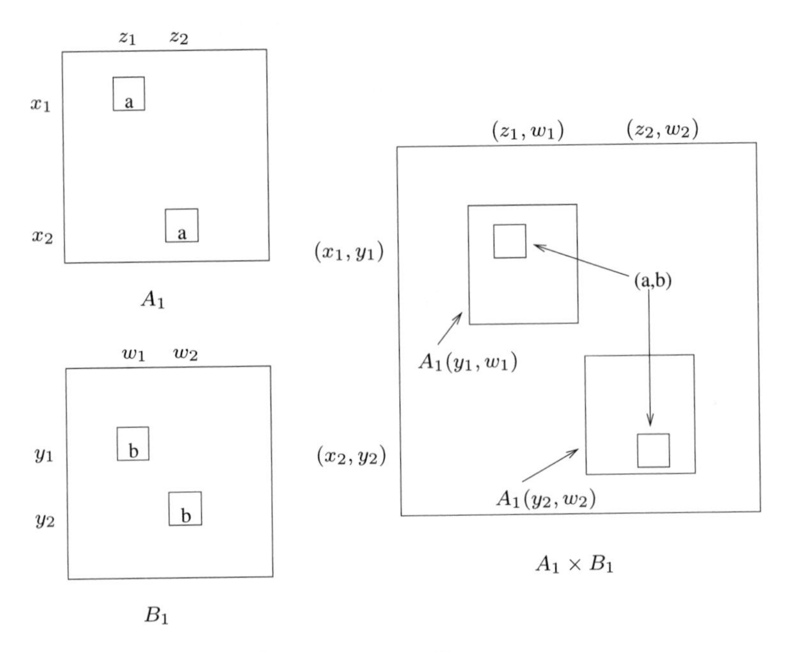

Cells (x_1, z_1) and (x_2, z_2) may be the same cell.

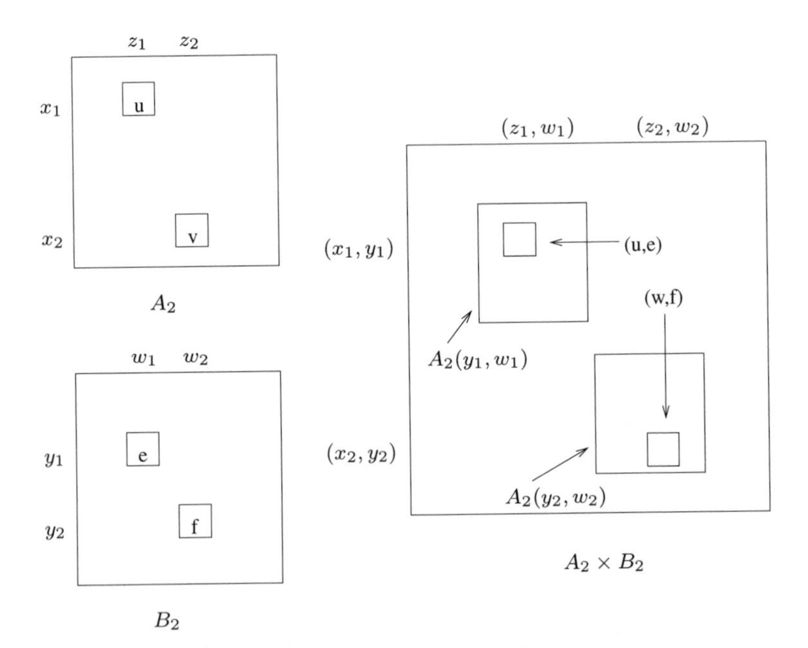

Figure 5.7: *The direct product of MOLS: case 2.*

□

Corollary 5.2.16 *If $L_1, L_2, \ldots L_t$ are t MOLS(m) and $M_1, M_2, \ldots, M_t$ are t MOLS(n), then the direct products $L_1 \times M_1$, $L_2 \times M_2$, $L_3 \times M_3, \cdots, L_t \times M_t$ are t MOLS(mn).* □

Corollary 5.2.17 *Let $n = p_1^{r_1} p_2^{r_2} \ldots p_x^{r_x}$ where $p_1, p_2, \ldots, p_x$ are distinct primes. There exist $m(n) = \min\{p_1^{r_1}, p_2^{r_2} \ldots, p_x^{r_x}\} - 1$ MOLS(n).*

Proof The Finite Field Construction produces $p_i^{r_i} - 1$ MOLS($p_i^{r_i}$) for $1 \leq i \leq x$. Since $m(n) \leq p_i^{r_i} - 1$ for all $1 \leq i \leq x$, selecting $m(n)$ MOLS($p_i^{r_i}$) for $1 \leq i \leq x$ and repeatedly applying the direct product construction produces $m(n)$ MOLS(n). □

Corollary 5.2.17 establishes that the bound in the MacNeish Conjecture can be attained. However, it turns out that this bound can actually be surpassed, as we shall see in the next section.

Exercises

5.2.18 Let $L_1 = L$ and $M_1 = M$ in Example 5.2.13. Let

$L_2 =$

1	3	2
3	2	1
2	1	3

and, $M_2 =$

1	2	3	4
2	1	4	3
3	4	1	2
4	3	2	1

Notice that L_1 is orthogonal to L_2 and M_1 is orthogonal to M_2. Form $L_2 \times M_2$ and check that $L_1 \times M_1$ is orthogonal to $L_2 \times M_2$, as Theorem 5.2.15 guarantees.

5.2.19 For the following values of n, use Corollary 5.2.17 to decide how many MOLS(n) can be produced using the finite field and direct productions.

(a) $n = 36$

(b) $n = 45$

(c) $n = 81$

(d) $n = 91$

(e) $n = 315$

5.3 Disproof of the MacNeish Conjecture

In 1959 E. T. Parker [18] took the first step toward disproving the Euler Conjecture by disproving the MacNeish Conjecture. Since the Euler Conjecture is a special case of the MacNeish Conjecture, a disproof of the MacNeish Conjecture is certainly a step in the right direction. What Parker did was to construct 3 MOLS(21). Since $21 = 3 \cdot 7$ and $2 = \min\{3, 7\} - 1$, this result disproves the MacNeish Conjecture. Clearly Corollary 5.2.17 *cannot* be used to construct 3 MOLS(21), so a different construction will be necessary. The following construction has come to be known as the PBD Construction.

The PBD Construction. Let (P, B) be a PBD of order n. Suppose that for each block $b \in B$ there exists an idempotent quasigroup $(b, \circ(b))$. Define a binary operation "$\circ$" on P as follows:

(1) $x \circ x = x$, all $x \in P$; and

(2) if $x \neq y$, $x \circ y = x \circ (b) y$, where $x, y \in b \in B$.

Then $(P, \circ)$ is an idempotent quasigroup. (See Figure 5.8.)

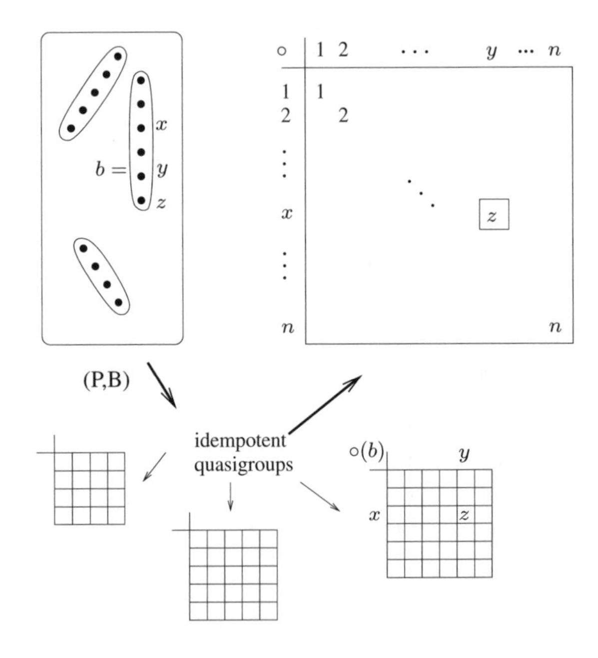

Figure 5.8: *The PBD Construction.*

Example 5.3.1 Let (P, B) be the PBD defined by $P = \{1, 2, 3, 4, 5, 6, 7\}$ and $B = \{\{1, 2, 4\}, \{2, 3, 5\}, \{3, 4, 6\}, \{4, 5, 7\}, \{5, 6, 1\}, \{6, 7, 2\}, \{7, 1, 3\}\}$. (So (P, B) is actually a Steiner triple system of order 7.)

$\circ_1$	1	2	4
1	1	4	2
2	4	2	1
4	2	1	4

$\circ_2$	2	3	5
2	2	5	3
3	5	3	2
4	3	2	5

$\circ_3$	3	4	6
3	3	6	4
4	6	4	3
6	4	3	6

$\circ_4$	4	5	7
4	4	7	5
5	7	5	4
7	5	4	7

$\circ_5$	5	6	1
5	5	1	6
6	1	6	5
1	6	5	1

$\circ_6$	6	7	2
6	6	2	7
7	2	7	6
2	7	6	2

$\circ_7$	7	1	3
7	7	3	1
1	3	1	7
3	1	7	3

$(P, \circ) =$

$\circ$	1	2	3	4	5	6	7
1	1	4	7	2	6	5	3
2	4	2	5	1	3	7	6
3	7	5	3	6	2	4	1
4	2	1	6	4	7	3	5
5	6	3	2	7	5	1	4
6	5	7	4	3	1	6	2
7	3	6	1	5	4	2	7

Theorem 5.3.2 *The PBD Construction does in fact produce an idempotent quasigroup.*

Proof Let (P, B) be a PBD and for each block $b \in B$ let $(b, \circ(b))$ be an idempotent quasigroup. Now regardless of whether (P, B) is a quasigroup, it is certainly idempotent. So suppose $x \circ y = x \circ z$. Then x, y, and z all belong to the same block b of B. Hence $x \circ (b) y = x \circ y = x \circ z = x \circ (b) z$ implies $y = z$. A similar argument shows that $y \circ x = z \circ x$ implies $y = z$. It follows that (P, B) is an idempotent quasigroup. □

Now the outstanding feature of the PBD Construction, as far as we are concerned, is that it can be used to construct orthogonal quasigroups.

Theorem 5.3.3 *Let (P, B) be a PBD with the property that for every block $b \in B$ there exists a pair of idempotent quasigroups $(b, \circ_1(b))$ and $(b, \circ_2(b))$ which are orthogonal. Then the PBD Construction produces a pair of idempotent* orthogonal *quasigroups $(P, \circ_1)$ and $(P, \circ_2)$.*

Proof We will use the Two Finger Rule. There are quite a few cases in the proof. We will handle just one here (the easiest one) and leave the remaining cases as an exercise. So suppose $x \circ_1 y = z \circ_1 w = u$, where $x \neq y$ and $z \neq w$. Let $x, y \in p$ and $z, w \in q$, where p and q are blocks belonging to B. Now since $(p, \circ_1(p))$ and $(p, \circ_2(p))$ are idempotent and orthogonal x, y, $x \circ_1 (p)y$, and $x \circ_2 (p)y$ are distinct. Similarly, z, w, $z \circ_1 (q)w$, and $z \circ_2 (q)w$ are distinct. If $x \circ_2 (p)y = z \circ_2 (q)w = v$, then $u \neq v$ belongs to both p and q and so $p = q$. But then $x \circ_1 (p)y = z \circ_1 (p)w$ *and* $x \circ_2 (p)y = z \circ_2 (p)w$ which contradicts the fact that $(p, \circ_1(p))$ and $(p, \circ_2(p))$ are orthogonal. As mentioned above, the other cases are left to Exercise 5.3.8. □

Corollary 5.3.4 *Let (P, B) be a PBD of order n and suppose that for each block $b \in B$ there exists m* idempotent *MOLS($|b|$). Then there exist m MOLS(n).* □

Theorem 5.3.5 *(Disproof of the MacNeish Conjecture.) There exist 3 MOLS(21)* [18].

Proof Let (P, B) be the following block design of order 21.

$$P = \{1, 2, 3, 4, 5, 6, 7, 8, 9, 10, 11, 12, 13, 14, 15, 16, 17, 18, 19, 20, 21\}$$

$$B = \left\{ \begin{array}{rrrrr|rrrrr} 21 & 1 & 2 & 3 & 4 & 2 & 6 & 10 & 14 & 17 \\ 21 & 5 & 6 & 7 & 8 & 2 & 7 & 9 & 16 & 19 \\ 21 & 9 & 10 & 11 & 12 & 2 & 8 & 11 & 13 & 20 \\ 21 & 13 & 14 & 15 & 16 & 3 & 5 & 10 & 16 & 20 \\ 21 & 17 & 18 & 10 & 20 & 3 & 6 & 12 & 13 & 19 \\ 1 & 5 & 9 & 13 & 17 & 3 & 7 & 11 & 15 & 17 \\ 1 & 6 & 11 & 16 & 18 & 3 & 8 & 9 & 14 & 18 \\ 1 & 7 & 12 & 14 & 20 & 4 & 5 & 11 & 14 & 19 \\ 1 & 8 & 10 & 15 & 19 & 4 & 6 & 9 & 15 & 20 \\ 2 & 5 & 12 & 15 & 18 & 4 & 7 & 10 & 13 & 18 \\ 4 & 8 & 12 & 16 & 17 & & & & & \end{array} \right.$$

Let $(Q, \circ_1)$, $(Q, \circ_2)$, and $(Q, \circ_3)$ be the three idempotent orthogonal quasigroups of order 5 given below:

$\circ_1$	x	y	z	u	v
x	x	u	y	v	z
y	u	y	v	z	x
z	y	v	z	x	u
u	v	z	x	u	y
v	z	x	u	y	v

$\circ_2$	x	y	z	u	v
x	x	z	v	y	u
y	v	y	u	x	z
z	u	x	z	v	y
u	z	v	y	u	x
v	y	u	x	z	v

$\circ_3$	x	y	z	u	v
x	x	v	u	z	y
y	z	y	x	v	u
z	v	u	z	y	x
u	y	x	v	u	z
v	u	z	y	x	v

For each block $\{x, y, z, u, v\} \in B$, where $x < y < z < u < v$, construct the 3 quasigroups given above.

Corollary 5.3.4 guarantees that the quasigroups $(P, \circ_1)$, $(P, \circ_2)$, and $(P, \circ_3)$ constructed using the PBD Construction are orthogonal. □

Exercises

5.3.6 Use the PBD of order 11 in Example 1.4.1 (1 block of size 5 and the remaining blocks of size 3) to construct an idempotent quasigroup of order 11.

5.3.7 Actually construct the 3 MOLS(21) in Theorem 5.3.5.

5.3.8 Supply *all* of the details in the proof of Theorem 5.3.3.

5.4 Disproof of the Euler Conjecture

Since the Euler Conjecture is a special case of the MacNeish Conjecture, once the MacNeish Conjecture has been shown to be false, there is no reason to believe that the Euler Conjecture is true. In 1959 Bose and Shrikhande [2] made the New York Times by constructing a pair of orthogonal latin squares of order 22.

We will give the original counter-example due to Bose and Shrikhande. We take this opportunity to point out that the PBD of order 22 in Lemma 4.2.1 with 1 block of size 7 and the remaining blocks of size 4 can be used to construct a pair of orthogonal latin squares of order 22 (using the PBD Construction). However, in 1959, the PBD in Lemma 4.2.1 had not as yet been constructed! (Hind sight is a wonderful thing.)

The example constructed by Bose and Shrikhande uses a slightly modified version of the PBD Construction. In what follows, a partial parallel class of blocks in a PBD will be called a *clear set* of blocks.

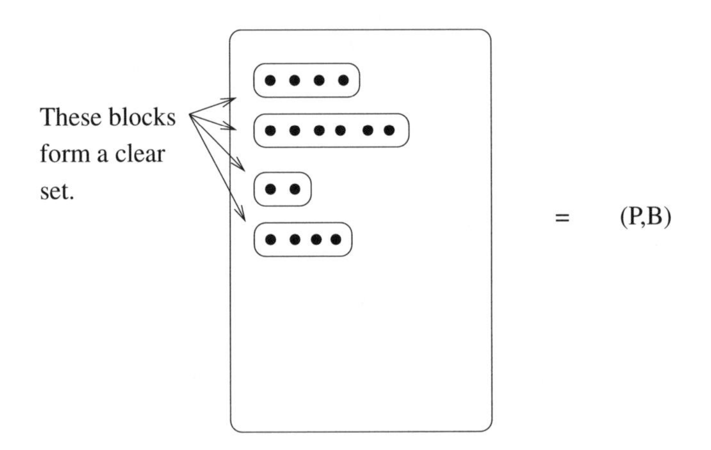

Figure 5.9: *Clear set of blocks.*

It is important to note that the blocks in a clear set need not have the same size nor be a complete parallel class; i.e., need not partition the points of the PBD.

The πPBD Construction. Let (P, B) be a PBD with a clear set $\pi \subseteq B$. For each block $p \in \pi$ let $(P, \circ(p))$ be *ANY* quasigroup, *NOT NECESSARILY IDEMPOTENT*. For each block $b \in B\backslash\pi$ let $(b, \circ(b))$ be an *idempotent* quasigroup. Now define a binary operation "$\circ$" on P by:

(1) if $x \in p \in \pi$, $x \circ x = x \circ (p)x$;

(2) if x does *not* belong to a block of the clear set π, $x \circ x = x$; and

(3) if $x \neq y \in b$, $x \circ y = x \circ (b)y$, *regardless* of whether $b \in \pi$ or $B\backslash\pi$.

Theorem 5.4.1 *The πPBD Construction does in fact produce a quasigroup.*

Proof The proof of this theorem is similar to the proof of Theorem 5.3.2 (with obvious modifications) and is left to Exercise 5.4.7. □

Theorem 5.4.2 *Let (P, B) be a PBD of order n with a clear set of blocks π such that for every block $p \in \pi$ there exists a pair of* not necessarily idempotent *orthogonal quasigroups of order $|p|$* and *for every block $b \in B \backslash \pi$ there exists a pair of* idempotent *orthogonal quasigroups of order $|b|$. Then the πPBD Construction produces a pair of orthogonal quasigroup of order n.* □

Proof The proof is similar to the proof of Theorem 5.3.3 (with obvious modifications) and is left to Exercise 5.4.7. □

Corollary 5.4.3 *Let (P, B) be a PBD of order n with a clear set of blocks π such that for each block p belonging to π there exists m MOLS ($|p|$) and for each block $b \in B \backslash \pi$ there exist m* idempotent *MOLS($|b|$). Then there exist m MOLS(n).* □

Theorem 5.4.4 *(Disproof of the Euler Conjecture.) There exist 2 MOLS(22)* [2].

Proof Let (P, B) be the following PBD of order 22. $P = \{1, \ldots, 22\}$, $\pi = \{\{1, 8, 12\}, \{5, 10, 15\}, \{20, 21, 22\}\}$, and

$$B = \left\{ \begin{array}{rrrrcrrrrr} 20 & 21 & 22 & & \quad & 1 & 2 & 3 & 4 & 5 \\ 1 & 8 & 12 & & & 6 & 7 & 8 & 9 & 10 \\ 5 & 10 & 15 & & & 11 & 12 & 13 & 14 & 15 \\ 16 & 17 & 18 & 19 & & 1 & 10 & 13 & 19 & 22 \\ 1 & 7 & 14 & 18 & & 1 & 6 & 11 & 16 & 21 \\ 2 & 6 & 14 & 20 & & 1 & 9 & 15 & 17 & 20 \\ 2 & 9 & 13 & 16 & & 2 & 7 & 12 & 17 & 22 \\ 2 & 10 & 11 & 18 & & 2 & 8 & 15 & 19 & 21 \\ 3 & 6 & 12 & 19 & & 3 & 8 & 13 & 18 & 20 \\ 3 & 7 & 15 & 16 & & 3 & 10 & 14 & 17 & 21 \\ 3 & 9 & 11 & 22 & & 4 & 6 & 15 & 18 & 22 \\ 4 & 7 & 13 & 21 & & 4 & 10 & 12 & 16 & 20 \\ 4 & 8 & 11 & 17 & & 5 & 7 & 11 & 19 & 20 \\ 4 & 9 & 14 & 19 & & 5 & 8 & 14 & 16 & 22 \\ 5 & 6 & 13 & 17 & & 5 & 9 & 12 & 18 & 21 \end{array} \right.$$

(Notice that the blocks in π in B form a clear set.) For each block belonging to B construct the appropriate pair of orthogonal quasigroups given below:

$\circ_1$	x	y	z
x	x	y	z
y	z	x	y
z	y	z	x

$\circ_2$	x	y	z
x	x	z	y
y	z	y	x
z	y	x	z

$x < y < z$

$\circ_1$	x	y	z	u
x	x	z	u	y
y	u	y	x	z
z	y	u	z	x
u	z	x	y	u

$\circ_2$	x	y	z	u
x	x	u	y	z
y	z	y	u	x
z	u	x	z	y
u	y	z	x	u

$$x < y < z < u$$

$\circ_1$	x	y	z	u	v
x	x	z	v	y	u
y	v	y	u	x	z
z	u	x	z	v	y
u	z	v	y	u	x
v	y	u	x	z	v

$\circ_2$	x	y	z	u	v
x	x	v	u	z	y
y	z	y	x	v	u
z	v	u	z	y	x
u	y	x	v	u	z
v	u	z	y	x	v

$$x < y < z < u < v$$

Theorem 5.4.2 guarantees that the quasigroups $(P, \circ_1)$ and $(P, \circ_2)$ constructed using the πPBD Construction are orthogonal. □

Exercises

5.4.5 Actually construct the pair of orthogonal latin squares of order 22 in Theorem 5.4.4.

5.4.6 Use the PBD of order 22 in Lemma 4.2.1 (1 block of size 7 and the remaining blocks of size 4) to give an alternate disproof of the Euler Conjecture.

5.4.7 Supply the proofs of Theorems 5.4.1 and 5.4.2.

5.5 Orthogonal latin squares of order $n \equiv 2$ (mod 4)

The Bose and Shrikhande Construction of a pair of orthogonal latin squares of order 22 kills the Euler Conjecture. It turns out that $n = 2$ and 6 are the *only* cases for which a pair of orthogonal latin squares fail to exist. Corollary 5.2.17

establishes the existence of a pair of orthogonal latin squares of every order $n \equiv 0, 1$ or $3 \pmod 4$. So to show that there exist a pair of orthogonal latin squares of every order $n \neq 2$ or 6, we need to handle the cases where $n \equiv 2 \pmod 4 \geq 10$. The cases $n = 10$ and 14 are handled by example. (See Examples 5.5.10 and 5.5.11.) The remaining cases, $n \equiv 2 \pmod 4 \geq 18$, will be constructed using a construction due to Richard Wilson. Actually, Wilson's Construction constructs a pair of orthogonal latin squares for every even $n \geq 18$ (and not just for $n \equiv 2 \pmod 4$).

A *transversal design* is a GDD (P, G, B) with all groups of the *same size* and such that each block intersects each group. We will denote a transversal design with group size m and block size n by $TD(m, n)$. Clearly the *order* of a transversal design $TD(m, n)$ is mn.

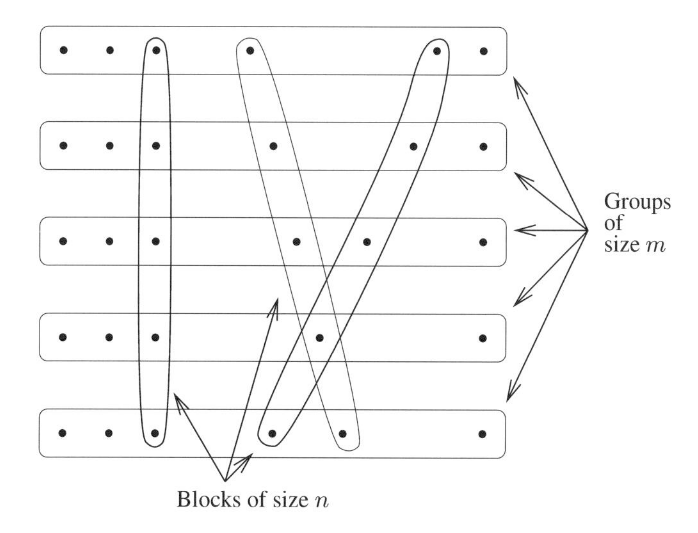

Figure 5.10: $TD(m, n)$.

Example 5.5.1

(a) $TD(3, 4)$

$G =$

1	2	3
4	5	6
7	8	9
10	11	12

$B =$

1	4	7	11	2	6	7	10
1	5	9	10	3	4	8	10
1	6	8	12	3	5	7	12
2	4	9	12	3	6	9	11
2	5	8	11				

(b) $TD(4, 3)$

$G =$

1	2	3	4
5	6	7	8
9	10	11	12

$B =$

1	5	9	3	5	12
1	6	10	3	6	9
1	7	11	3	7	10
1	8	12	3	8	11
2	5	10	4	5	11
2	6	11	4	6	12
2	7	12	4	7	9
2	8	9	4	8	10

(c) $TD(3, 3)$

$G =$

1	5	9
2	6	7
3	4	8

$B =$

1	2	3	3	5	7
4	5	6	1	4	7
7	8	9	2	5	8
1	6	8	3	6	9
2	4	9			

A set of k mutually orthogonal quasigroups (latin squares) of order n is *equivalent* to a $TD(n, k+2)$. This is quite easy to see. Let $(L, \circ_1), (L, \circ_2), \ldots, (L, \circ_k)$ be k mutually orthogonal quasigroups of order n. Set

$$\begin{cases} P &= \{1, 2, 3, \ldots, n\} \times \{1, 2, \ldots, k+2\}, \\ G &= \{\{(1, i), (2, i), (3, i), \ldots, (n, i)\} \mid i = 1, 2, \ldots, k+2\}, \text{ and} \\ B &= \{\{(i, 1), (j, 2), (i \circ_1 j, 3), (i \circ_2 j, 4), \ldots, (i \circ_k j, k+2)\} \mid \text{all } i, \\ & \qquad j \in \{1, 2, 3, \ldots, n\}\}. \end{cases}$$

It is straight forward and not difficult to see that (P, G, B) is a $TD(n, n+2)$. Conversely, suppose we have the $TD(n, k+2)$ given by

$$\begin{cases} P &= \{1, 2, 3, \ldots, n\} \times \{1, 2, 3, \ldots, k+2\}, \\ G &= \{\{(i, i), (2, i), (3, i), \ldots, (n, i)\} \mid i = 1, 2, \ldots, k+2\}, \text{ and} \\ n^2 & \text{blocks } B. \end{cases}$$

Define k *groupoids* $(L, \circ_1), (L, \circ_2), \ldots, (L, \circ_k)$ of order n by: $x \circ_i y = z_i$, if and only if

$$\{(x, 1), (y, 2), (z_1, 3), (z_2, 4), \ldots, (z_k, k+2)\} \in B.$$

Once again it is easy to see that the k groupoids $(L, \circ_1), (L, \circ_2), \ldots, (L, \circ_k)$ are mutually orthogonal *quasigroups*. (The proof that a $TD(n, k+2)$ is equivalent to k mutually orthogonal quasigroups of order n is left to Exercise 5.5.3.)

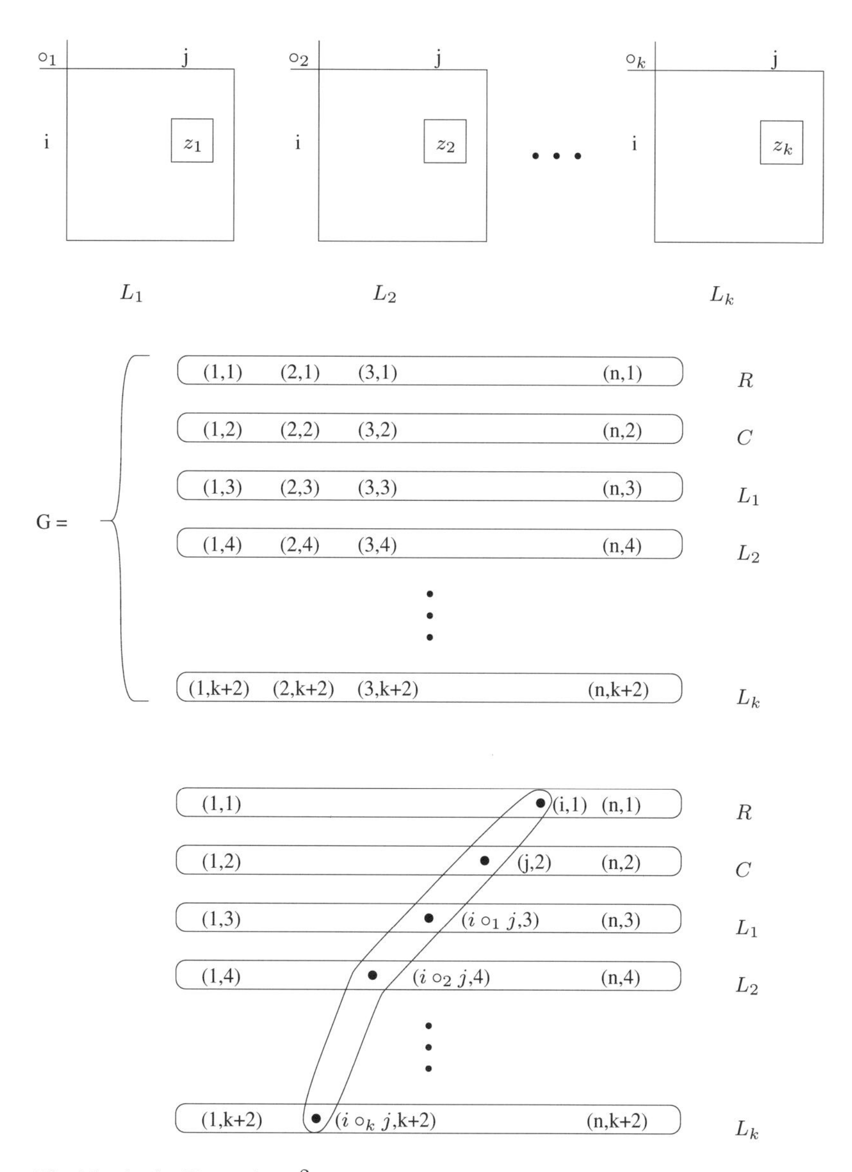

The blocks in B are the n^2 sets
$\{(i,1),(j,2),(i \circ_1 j,3),(i \circ_2 j,4),...,(i \circ_k j,k+2)\}$,
for all $i,j \in \{1,2,...,n\}$.

Figure 5.11: $TD(n,k+2) \equiv k$ *MOLS*(n).

Example 5.5.2

$\circ_1$	1	2	3
1	1	3	2
2	3	2	1
3	2	1	3

L_1

$\circ_2$	1	2	3
1	1	2	3
2	3	1	2
3	2	3	1

L_2

G =

11	21	31	= R
12	22	32	= C
13	23	33	= L_1
14	24	24	= L_2

$B =$

11	12	13	14	21	32	13	24
11	22	33	24	31	12	23	23
11	32	23	34	31	32	33	13
21	12	33	34	31	22	13	33
21	22	23	14				

Exercises

5.5.3 Prove that a $TD(n, k+2)$ is equivalent to k mutually orthogonal quasigroups of order n.

5.5.4 Construct a $TD(4, 4)$.

(The above $TD(4, 4)$ and the $TD(3, 4)$ in Example 5.5.2 turn out to be quite useful in what follows.)

We now give a series of constructions leading up to the main construction. The reason for doing this, is that it makes the main construction a lot easier to swallow.

The m $\rightarrow$ 3m Construction (If a $TD(m, 4)$ exists, then a $TD(3m, 4)$ exists.)

Let (P, G, B) be a $TD(m, 4)$ and let $W = \{1, 2, 3\}$. Let $P^* = W \times P$ and define a collection of groups G^* and blocks B^* as follows:

(1) $G^* = \{W \times g \mid g \in G\}$, and

(2) for each block $b \in B$, let $(W \times b, \{W \times \{a\} \mid a \in b\}, W(b))$ be a $TD(3, 4)$, and place the 9 blocks belonging to $W(b)$ in B^*.

Then (P^*, G^*, B^*) is a $TD(3m, 4)$. See Figure 5.12.

(It is worth remarking that the $m \rightarrow 3m$ Construction is just the direct product in design theory clothing.)

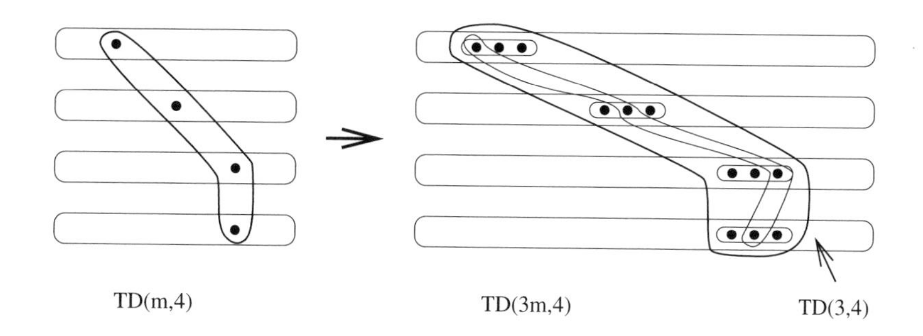

Figure 5.12: *$m \to 3m$ Construction.*

Exercises

5.5.5 Construct a $TD(12,4)$ from a $TD(4,4)$.

The m $\to$ 3m+1 Construction (If a $TD(m,5)$ exists, then a $TD(3m+1,4)$ exists.)

Let (P,G,B) be a $TD(m,5)$ with groups $G = \{g_1,g_2,g_3,g_4,g_5\}$. Let $x \in g_5$ and set $\pi = \{b\backslash\{x\} \mid x \in b \in B\}$. Define $P_1 = P\backslash g_5$, $G_1 = \{g_1,g_2,g_3,g_4\}$, and $B_1 = \{b\backslash g_5 \mid b \in B\}$. Then (P_1,G_1,B_1) is a $TD(m,4)$ and π is a *parallel class* of blocks of B_1. We now define a $TD(3m+1,4)$ as follows: Let $W = \{1,2,3\}$ and set $P_2 = \{\infty_1,\infty_2,\infty_3,\infty_4\} \cup (W \times P_1)$. Define a collection of groups G_2 and blocks B_2 as follows:

(1) $G_2 = \{\{\infty_i\} \cup (W \times g_i) \mid g_i \in G_1\}$,

(2) for each block $p \in \pi$, let $(\{\infty_1,\infty_2,\infty_3,\infty_4\} \cup (W \times p), \{\{\infty_i\} \cup (W \times \{a\}) \mid a \in p \cap g_i, i \in W\}, W(p))$ be a $TD(4,4)$ with the proviso that $\{\infty_1,\infty_2,\infty_3,\infty_4\}$ is a block, and place the 15 blocks of $W(p)\backslash\{\{\infty_1,\infty_2,\infty_3,\infty_4\}\}$ in B_2,

(3) for each block $b \in B_1\backslash\pi$, let $(W \times b, \{W \times \{a\} \mid a \in b\}, W(b))$ be a $TD(3,4)$ and place the 9 blocks of $W(b)$ in B_2, and

(4) $\{\infty_1,\infty_2,\infty_3,\infty_4\} \in B_2$.

Then (P_2,G_2,B_2) is a $TD(3m+1,4)$. See Figure 5.13.

Exercises

5.5.6 Construct a $TD(13,4)$ from a $TD(4,5)$.

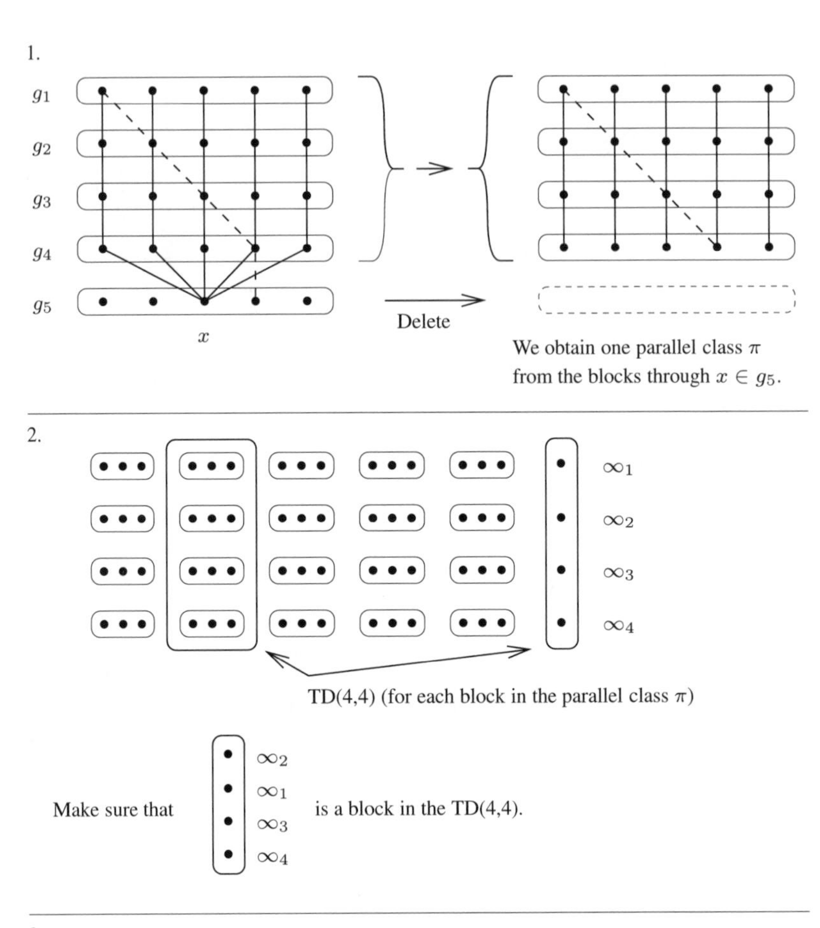

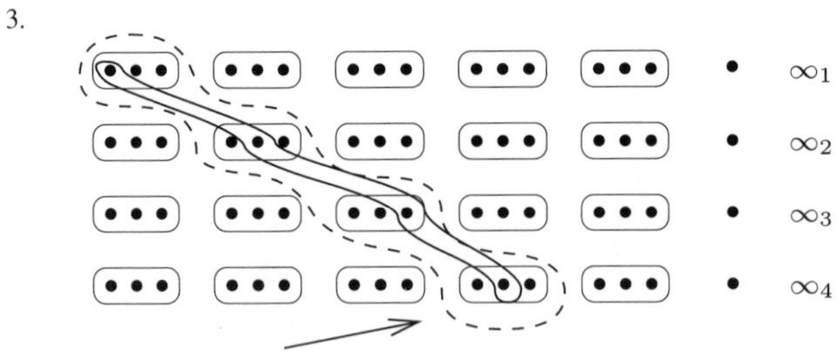

4.

Finally, include the block $\{\infty_1, \infty_2, \infty_3, \infty_4\}$ exactly once.

Figure 5.13: *The $m \to 3m + 1$ Construction.*

We now give THE CONSTRUCTION!

The m → 3m+u Construction (If a $TD(m,5)$ exists *and* a $TD(u,4)$ exists where $u \leq m$, then a $TD(3m+u,4)$ exists.)

Let (P,G,B) be a $TD(m,5)$ with groups $G = \{g_1, g_2, g_3, g_4, g_5\}$. Let $U = \{x_1, x_2, \ldots, x_u\} \subseteq g_5$ and set $\pi_i = \{b \backslash \{x_i\} \mid x_i \in b \in B\}$. Define $P_1 = P \backslash g_5$, $G_1 = \{g_1, g_2, g_3, g_4\}$, and $B_1 = \{b \backslash g_5 \mid b \in B\}$. Then (P_1, G_1, B_1) is a $TD(m,4)$ and $\pi_1, \pi_2, \ldots, \pi_u$ are *parallel classes* of blocks of B_1. Define a $TD(3m+u,4)$ as follows: Let $W = \{1,2,3\}$, $U^* = U \times \{1,2,3,4\}$, and set $P_2 = (W \times P_1) \cup (U \times \{1,2,3,4\})$. Define a collection of groups G_2 and blocks B_2 as follows:

(1) $G_2 = \{(U \times \{i\}) \cup (w \times g_i) \mid g_i \in G_1\}$;

(2) for each block $p = \{a,b,c,d\} \in \pi_i$, let $((\{x_i\} \times \{1,2,3,4\}) \cup (W \times p)$, G^*, $W(p))$ be a $TD(4,4)$ where $G^* = \{(x_i, j)\} \cup (W \times \{a\}) \mid a \in p \cap g_j \in G_1\}$, and $\{(x_i,1),(x_i,2),(x_i,3),(x_i,4)\} \in W(p)$, and place the 15 blocks of $W(p) \backslash \{(x_i,1),(x_i,2),(x_i,3),(x_i,4)\}$ in B_2,

(3) for each block $b \in B_1 \backslash \pi_1 \cup \pi_2 \cup \cdots \cup \pi_u$, let $(W \times b, \{W \times \{a\} \mid a \in b\}, W(b))$ be a $TD(3,4)$, and place the 9 blocks of $W(b)$ in B_2, and

(4) let $(U \times \{1,2,3,4\}, \{U \times \{i\} \mid i = 1,2,3,4\}, B^*)$ be a $TD(u,4)$ and place the u^2 blocks of B^* in B_2.

Then (P_2, G_2, B_2) is a $TD(3m+u,4)$. See Figure 5.14.

Exercises

5.5.7 Construct a $TD(18,4)$ from a $TD(5,5)$ and a $TD(3,4)$.

5.5.8 Let (P,G,B) be a triple where P is a set of size mn, G is a collection of n pairwise disjoint sets of size m which partition P, and B is a collection of sets of size n each of which intersects each set in G. Prove that if each pair of points in different sets of G belong to at least one set of B and $|B| \leq m^2$, then (P,G,B) is a $TD(m,n)$. (This is a slick way to prove that a triple (P,G,B) is a $TD(m,n)$.)

5.5.9 Prove that The $m \to 3m+u$ Construction gives a $TD(3m+u,4)$.

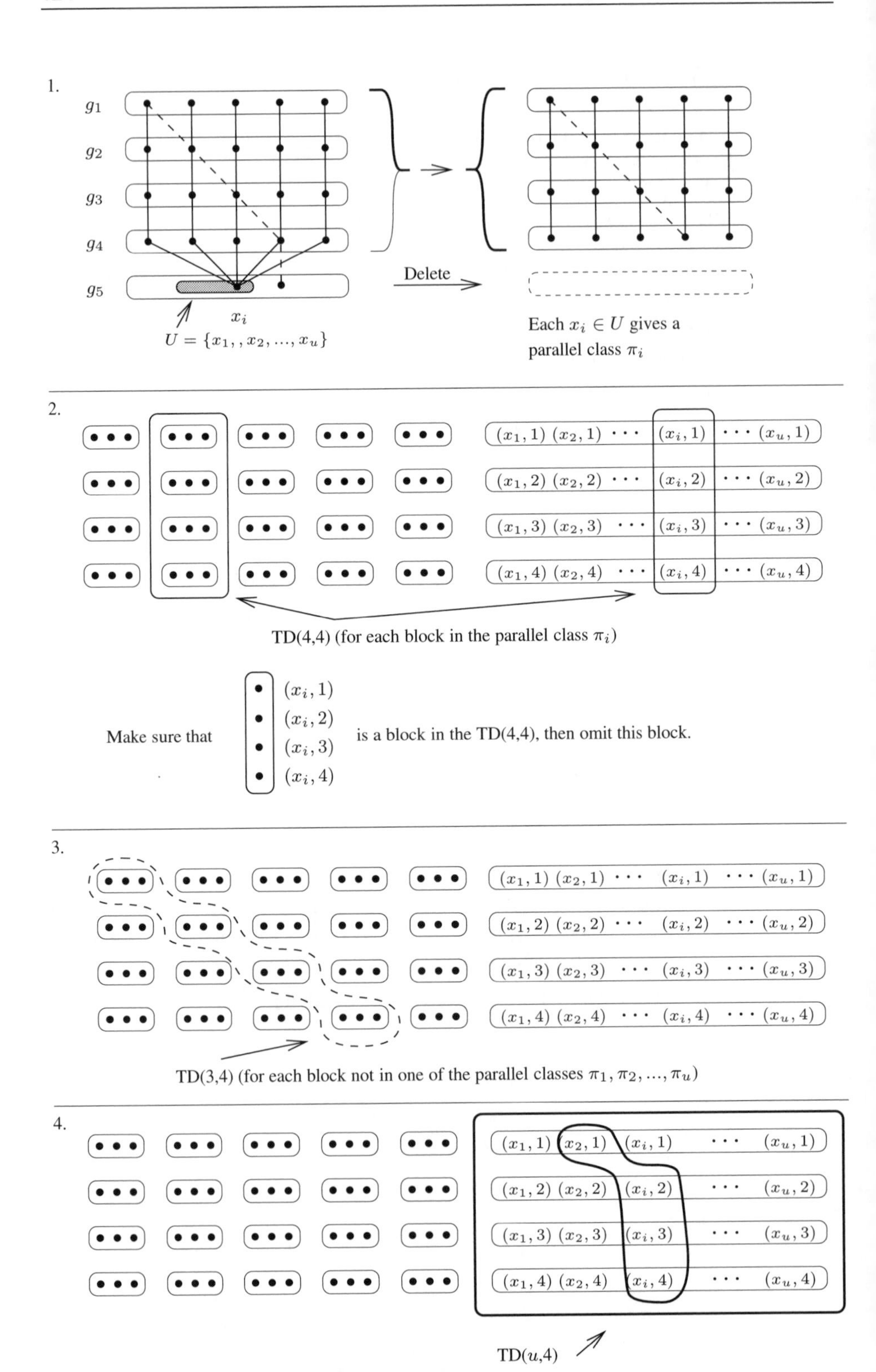

Figure 5.14: *The $m \to 3m + u$ Construction.*

Example 5.5.10 (2 MOLS(10)).

4	10	9	8	3	2	7	5	6	1
2	3	7	5	4	10	9	8	1	6
8	1	6	9	10	4	5	3	2	7
9	8	1	4	5	6	3	2	7	10
10	9	8	6	1	3	2	7	4	5
7	2	3	1	6	5	4	10	9	8
5	4	10	3	2	7	6	1	8	9
6	5	4	2	7	1	8	9	10	3
1	6	5	7	8	9	10	4	3	2
3	7	2	10	9	8	1	6	5	4

5	4	10	1	2	7	8	9	3	6
3	1	6	4	8	5	9	2	10	7
10	9	8	7	3	6	1	4	5	2
2	5	4	3	6	1	7	8	9	10
9	8	7	6	1	10	4	5	2	3
1	6	3	5	9	2	10	7	4	8
8	7	2	9	10	4	5	3	6	1
4	10	9	2	7	8	3	6	1	5
7	2	5	10	4	3	6	1	8	9
6	3	1	8	5	9	2	10	7	4

Example 5.5.11 (2 MOLS(14)).

12	7	14	6	4	2	11	9	13	5	3	1	10	8
1	11	10	4	5	6	13	8	9	14	7	12	2	3
2	1	11	14	7	12	3	4	5	6	13	8	9	10
10	9	8	1	2	3	4	5	6	13	14	7	12	11
7	14	12	3	1	10	8	6	4	2	11	9	13	5
3	2	1	13	8	9	10	11	14	7	12	4	5	6
5	4	3	10	11	1	2	14	7	12	6	13	8	9
13	6	5	2	3	4	14	7	12	8	9	10	11	1
9	8	13	5	6	14	7	12	10	11	1	2	3	4
11	10	9	8	14	7	12	1	2	3	4	5	6	13
14	12	7	11	9	13	5	3	1	10	8	6	4	2
4	3	2	7	12	5	6	13	8	9	10	11	1	14
6	5	4	12	13	8	9	10	11	1	2	3	14	7
8	13	6	9	10	11	1	2	3	4	5	14	7	12

12	13	4	10	7	14	1	9	6	3	11	8	5	2
9	8	10	5	6	7	4	13	12	11	1	2	3	14
5	14	6	7	8	9	10	11	1	2	3	4	13	12
6	5	7	1	2	3	14	4	13	12	8	9	10	11
4	12	13	6	3	11	8	5	2	10	7	14	1	9
1	11	2	9	10	4	13	12	3	14	5	6	7	8
14	3	5	2	4	13	12	6	7	8	9	10	11	1
7	6	8	4	13	12	9	10	11	1	2	3	14	5
10	9	11	13	12	1	2	3	14	5	6	7	8	4
2	1	3	12	14	5	6	7	8	9	10	11	4	13
13	4	12	3	11	8	5	2	10	7	14	1	9	6
8	7	9	11	1	2	3	14	5	6	4	13	12	10
11	10	1	14	5	6	7	8	9	4	13	12	2	3
3	2	14	8	9	10	11	1	4	13	12	5	6	7

With the $m \to 3m + u$ Construction and the two examples given above we are ready to drive a stake through Euler's coffin.

Theorem 5.5.12 *There exist a pair of mutually orthogonal latin squares of order n for every $n \neq 2$ or 6.*

Proof We already know that there does *not* exist a pair of MOLS(n) for $n = 2$ and 6, and Corollary 5.2.17 produces a pair of MOLS(n) for every $n \equiv 0, 1,$ or 3 (mod 4). So we need to construct a pair of MOLS(n) for every $n \equiv 2$ (mod 4) ≥ 10. Examples 5.5.10 and 5.5.11 take care of 10 and 14, so we need handle only the cases $n \equiv 2$ (mod 4) ≥ 18. Define $n, m,$ and u as in the accompanying table.

n	m	u
$18k$	$6k-1$	3
$18k+2$	$6k-1$	5
$18k+4$	$6k+1$	1
$18k+6$	$6k+1$	3
$18k+8$	$6k+1$	5
$18k+10$	$6k+1$	7
$18k+12$	$6k+1$	9
$18k+14$	$6k+1$	11
$18k+16$	$6k+5$	1

Then a $TD(m, 5)$ and a $TD(u, 4)$ exist for *every* $k \geq 1$, except $30 = 18 \cdot 1 + 12$ *and* $32 = 18 \cdot 1 + 14$. (See Exercise 5.5.13.) Hence the $m \to 3m + u$ Construction handles everything except 30 and 32. However, the Finite Field Construction takes care of 32 and since there exists a pair of MOLS(10), the direct product takes care of 30. □

Exercises

5.5.13 Show that for every $k \geq 1$, there exists a $TD(6k-1, 5)$ and there exists a $TD(6k+1, 5)$. (Hint, What is the smallest prime that can possibly divide $6k - 1$ or $6k + 1$?)

5.5.14 Let $A, B,$ and C be the three MOLS(7) given below.

1	6	4	2	7	5	3
4	2	7	5	3	1	6
7	5	3	1	6	4	2
3	1	6	4	2	7	5
6	4	2	7	5	3	1
2	7	5	3	1	6	4
5	3	1	6	4	2	7

A

1	4	7	3	6	2	5
6	2	5	1	4	7	3
4	7	3	6	2	5	1
2	5	1	4	7	3	6
7	3	6	2	5	1	4
5	1	4	7	3	6	2
3	6	2	5	1	4	7

B

1	2	3	4	5	6	7
7	1	2	3	4	5	6
6	7	1	2	3	4	5
5	6	7	1	2	3	4
4	5	6	7	1	2	3
3	4	5	6	7	1	2
2	3	4	5	6	7	1

C

For each $x \in \{1,2,3,4,5,6,7\}$ denote by A_x the transversal of A determined by x in C and by B_x the transversal of B determined by x in C. For example, $A_2 = \{(1,2,6),(2,3,7),(3,4,1),(4,5,2),(5,6,3),(6,7,4),(7,1,5)\}$ and $B_2 = \{(1,2,4),(2,3,5),(3,4,6),(4,5,7),(5,6,1),(6,7,2),(7,1,3)\}$, where (x,y,z) means that cell (x,y) is occupied by z. Find a pair of DISJOINT 3-element subsets $\{x_1,x_2,x_3\}$ and $\{y_1,y_2,y_3\}$ of $\{1,2,3,4,5,6,7\}$ and projections of $A_{x_1}, A_{x_2}, A_{x_3}$ and $B_{y_1}, B_{y_2}, B_{y_3}$ so that the partial latin squares below can be completed to a pair of orthogonal latin squares of order 10.

8	10	9							
10	9	8							
9	8	10							
			1	6	4	2	7	5	3
			4	2	7	5	3	1	6
			7	5	3	1	6	4	2
			3	1	6	4	2	7	5
			6	4	2	7	5	3	1
			2	7	5	3	1	6	4
			5	3	1	6	4	2	7

9	10	8							
8	9	10							
10	8	9							
			1	4	7	3	6	2	5
			6	2	5	1	4	7	3
			4	7	3	6	2	5	1
			2	5	1	4	7	3	6
			7	3	6	2	5	1	4
			5	1	4	7	3	6	2
			3	6	2	5	1	4	7

5.5.15 Construct 3 mutually orthogonal latin squares of order 11 and use the above technique to construct a pair of orthogonal latin squares of order 14.

Chapter 6

Affine and Projective Planes

6.1 Affine planes.

In what *follows* we will call the blocks of a PBD *lines*. If several points belong to the same line we will say that they are *collinear*, and if two lines fail to intersect we will say that they are *parallel*. Additionally, if the point p belongs to the line ℓ we will say that p is *on* the line and that ℓ *passes through* p.

An *affine plane* is a PBD (P, B) with the following properties:

(1) P contains at least one subset of 4 points no 3 of which are collinear, and

(2) (*parallel postulate*) given a line ℓ and a point p not on ℓ, there is EXACTLY one line of B containing p which is parallel to ℓ.

(1)

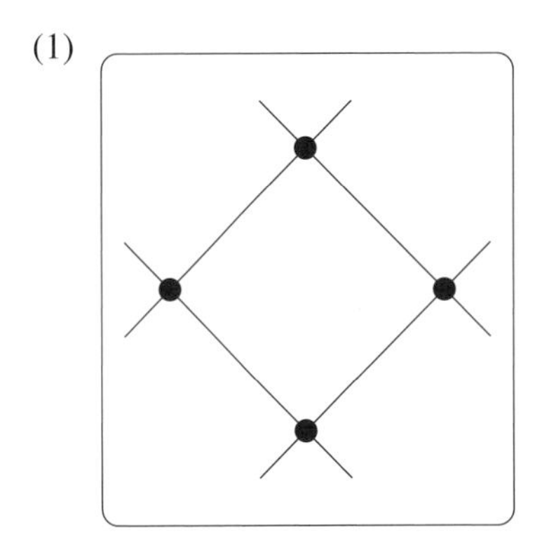

4 points, no 3 of which are collinear.

(2)

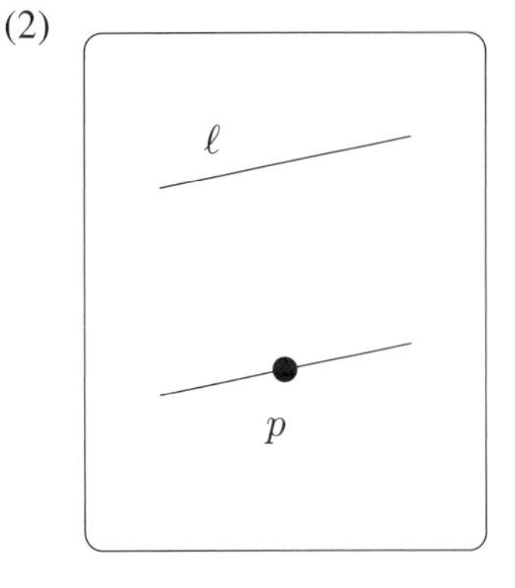

There is exactly one line through p parallel to ℓ.

Figure 6.1: *Affine plane.*

Remark Condition (1) in the definition of an affine plane guarantees that the plane is *non-trivial*; that is, does not consist of a single line.

Example 6.1.1

(a)
$$P = \{1, 2, 3, 4\}$$
$$B = \left\{ \begin{array}{lll} \{1,2\} & \{1,3\} & \{1,4\} \\ \{3,4\} & \{2,4\} & \{2,3\} \end{array} \right.$$

(b)
$$P = \{1, 2, 3, 4, 5, 6, 7, 8, 9\}$$
$$B = \left\{ \begin{array}{llll} \{1,2,3\} & \{1,4,7\} & \{1,5,9\} & \{1,6,8\} \\ \{4,5,6\} & \{2,5,8\} & \{2,6,7\} & \{2,4,9\} \\ \{7,8,9\} & \{3,6,9\} & \{3,4,8\} & \{3,5,7\} \end{array} \right.$$

In Example 6.1.1(a) there is, clearly, just one collection of 4 points, no 3 of which are collinear. However, in general, such collections are far from unique. For example, in Example 6.1.1(b) there are exactly 54 collections of 4 points, no 3 of which are collinear: $\{1, 2, 6, 9\}$ and $\{3, 5, 6, 8\}$ are 2 such collections.

Exercises

6.1.2 Prove that in the affine plane in Example 6.1.1(b) there are *exactly* 54 collections of 4 points, no 3 of which are collinear.

6.1.3 Let (P, B) be an affine plane. If $n \geq 2$, prove that the following statements are equivalent:

(a) One line contains n points.
(b) One point belongs to exactly $n + 1$ lines.
(c) Every line contains n points.
(d) Every point is on exactly $n + 1$ lines.
(e) There are exactly n^2 points in P.
(f) There are exactly $n^2 + n$ lines in B.

The *number* n is called the *order* of the affine plane (P, B); i.e., the *number* of points on each line is called the order of the affine plane.

Exercise 6.1.3 shows that an affine plane is simply a block design containing n^2 points with each block containing n points. The *converse* is also true; i.e., a block design containing n^2 points with block size n is an affine plane.

Exercises

6.1.4 Show that a block design of order n^2 with block size n is an affine plane.

Remark *Unfortunately*, if (P, B) is a block design with $|P| = n^2$ points and block size n, the *order* is n^2 if (P, B) is considered as a block design, and is n if (P, B) is considered as an affine plane. So the word "order" can mean two different things. The reason for this is that affine planes were studied long before

PBDs and block designs. While the order of an affine plane determines the number of points, clearly this is *not* the case for block designs in general. That is to say, the block size of a block design does not, in general, determine the number of points. We just need to pay attention to the context in which we use the word "order" if we are bouncing back and forth between PBDs and affine planes.

6.2 Projective planes.

A *projective plane* is a PBD (P, B) with the following properties:

(1) P contains at least one subset of 4 points, no 3 of which are collinear; and

(2) every pair of lines *intersect* in EXACTLY one point.

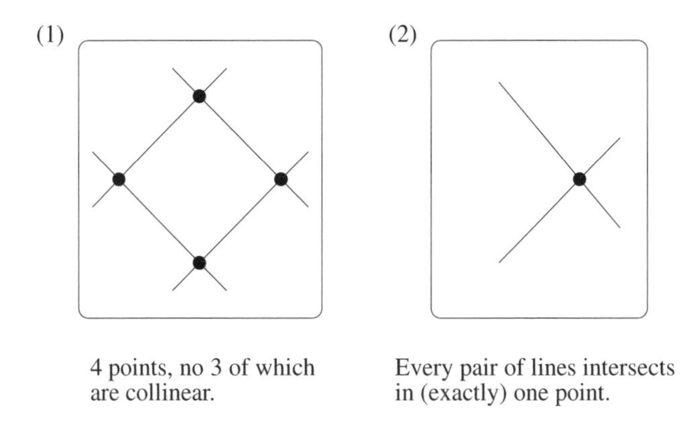

Figure 6.2: *Projective plane.*

Remark Condition (1) in the definition of a projective plane guarantees that the plane is *non-degenerate*; that is, does not consist of $n+1$ points, with one line of size n and the remaining lines of length 2 (= the *degenerate projective plane*).

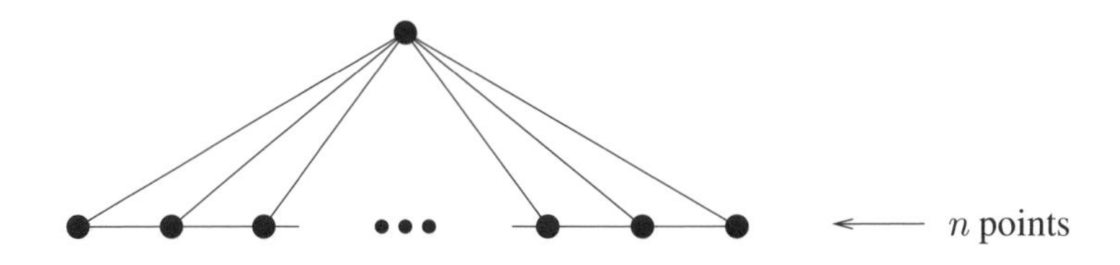

Figure 6.3: *Degenerate projective plane (*not *a projective plane according to the definition).*

Example 6.2.1

(a) $P^* = \{1,2,3,4,5,6,7\}$

$$B^* = \left\{ \begin{array}{llll} \{1,2,5\} & \{1,3,6\} & \{1,4,7\} & \{5,6,7\} \\ \{3,4,5\} & \{2,4,6\} & \{2,3,7\} & \end{array} \right.$$

(b) $P^* = \{1,2,3,4,5,6,7,8,9,10,11,12\}$

$$B^* = \left\{ \begin{array}{llll} \{1,2,3,10\} & \{1,4,7,11\} & \{1,5,9,12\} & \{1,6,8,13\} \\ \{4,5,6,10\} & \{2,5,8,11\} & \{2,6,7,12\} & \{2,4,9,13\} \\ \{7,8,9,10\} & \{3,6,9,11\} & \{3,4,8,12\} & \{3,5,7,13\} \\ \{10,11,12,13\} & & & \end{array} \right.$$

Exercises

6.2.2 Let (P, B) be a projective plane. If $n \geq 2$, the following statements are equivalent:

(a) One line contains $n + 1$ points.
(b) One point belongs to exactly $n + 1$ lines.
(c) Every line contains $n + 1$ points.
(d) Every point is on exactly $n + 1$ lines.
(e) There are exactly $n^2 + n + 1$ points in P.
(f) There are exactly $n^2 + n + 1$ lines in B.

The number n is called the *order* of the projective plane (P, B); i.e., the order of the projective plane is the number *one less* than the number of points on each line.

Exercise 6.2.2 shows that a projective plane is a block design containing $n^2 + n + 1$ points with each block containing $n + 1$ points. The *converse* is also true.

Exercises

6.2.3 Show that a block design of order $n^2 + n + 1$ with block size $n + 1$ is a projective plane.

Remark So that we can keep all of this straight: Let (P, B) be a block design. If (P, B) is considered as a block design, the order is $|P|$. If (P, B) is also an affine plane and is considered as an affine plane, the order is the *block size*. If (P, B) is also a projective plane and is considered as a projective plane, the order is *one less* than the block size.

6.3 Connections between affine and projective planes.

In an affine plane (P, B), a collection of mutually parallel lines which partition the points of P is called a *parallel class* (also known as a *pencil* of lines).

Exercises

6.3.1 In an affine plane, if a line intersects one of two parallel lines, then it also intersects the other.

6.3.2 An affine plane of order n has exactly $n+1$ parallel classes, each containing n lines.

6.3.3 Let (P, B) be an affine plane of order n and denote the $n+1$ parallel classes by $\pi_1, \pi_2, \pi_3, \ldots, \pi_{n+1}$. Let $\infty = \{\infty_1, \infty_2, \ldots, \infty_{n+1}\}$ be a set of $n+1$ *distinct* symbols, none of which belong to P. Set

$$\begin{cases} P(\infty) = P \cup \infty, \text{ and} \\ B(\infty) = \{b \cup \{\infty_i\} \mid b \in \pi_i\} \cup \{\infty\}. \end{cases}$$

Prove that

$$(P(\infty), B(\infty))$$

is a projective plane of order n. (The technique of constructing $(P(\infty), B(\infty))$ from (P, B) is called *adding a line at infinity*, and ∞ is called the *line at infinity*.)

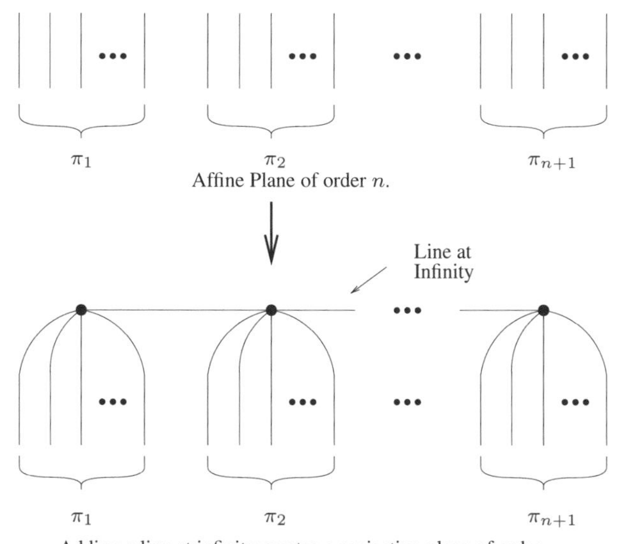

Affine Plane of order n.

Adding a line at infinity creates a projective plane of order n.

(The projective planes in Example 6.2.1 were obtained by adding a line at infinity to the affine planes in Example 6.1.1.)

6.3.4 Let (P, B) be the affine plane where
$P = \{1, 2, 3, 4, 5, 6, 7, 8, 9, 10, 11, 12, 13, 14, 15, 16\}$ and
$B = \{\{1, 2, 3, 4\}, \{1, 5, 9, 14\}, \{4, 5, 10, 13\}, \{2, 6, 10, 14\}, \{2, 7, 12, 13\}, \{3, 8, 10, 16\}, \{3, 5, 12, 15\}, \{5, 6, 7, 8\}, \{4, 7, 9, 16\}, \{4, 8, 12, 14\}, \{1, 6, 12, 16\}, \{1, 7, 10, 15\}, \{3, 6, 9, 13\}, \{2, 8, 9, 15\}, \{9, 10, 11, 12\}, \{3, 7, 11, 14\}, \{4, 6, 11, 15\}, \{13, 14, 15, 16\}, \{2, 5, 11, 16\}, \{1, 8, 11, 13\}\}$. Organize the lines into parallel classes and add the line $\{17, 18, 19, 20, 21\}$ at infinity to obtain a projective plane of order 4.

6.3.5 If (P, B) is any PBD and X is any subset, then the set $P \setminus X$ equipped with the set of subsets $b \setminus X$ for each $b \in B$ is a PBD and is said to be *derived* from (P, B) by deleting the points in X. If (P, B) is a projective plane of order n and ∞ is *any* line of B, prove that the block design derived from (P, B) by deleting the points on ∞ is an affine plane.

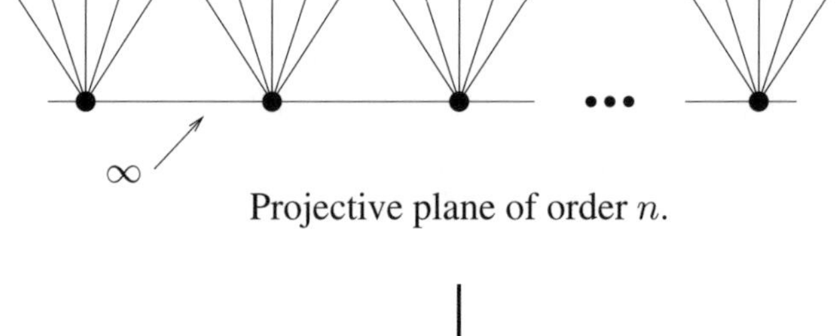

Projective plane of order n.

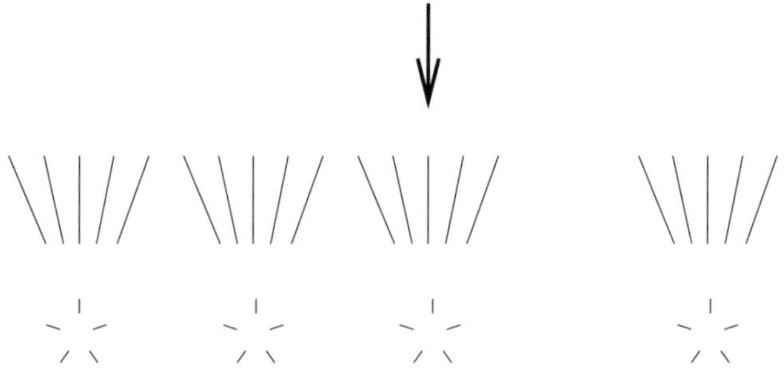

Affine plane formed by deleting the line ∞.

6.3.6 Let (P, B) be the projective plane of order 4, where

$$P = \{1, 2, 3, \ldots, 21\}, \text{ and}$$
$$B = \begin{cases} \{\{1,2,3,4,5\}, & \{3,7,11,14,18\}, & \{3,8,10,17,20\}, \\ \{5,9,13,17,18\}, & \{4,9,10,15,19\}, & \{1,14,15,16,17\}, \\ \{2,6,10,14,18\}, & \{5,6,12,15,20\}, & \{3,6,13,16,19\}, \\ \{2,9,11,16,20\}, & \{5,8,11,14,19\}, & \{5,7,10,16,21\}, \\ \{2,7,12,17,19\}, & \{1,6,7,8,9\}, & \{4,8,12,16,18\}, \\ \{1,10,11,12,13\}, & \{4,7,13,14,20\}, & \{3,9,12,14,21\}, \\ \{1,18,19,20,21\}, & \{2,8,13,15,21\}, & \{4,6,11,17,21\}\}. \end{cases}$$

Delete the line $\{2, 8, 13, 15, 21\}$ from (P, B) to obtain an affine plane of order 4. Be sure to arrange the resulting lines into 5 parallel classes.

Remark The symbiotic relationship between affine and projective planes explains why the order of a projective plane is *one less* than the size of the lines.

6.4 Connection between affine planes and complete sets of MOLS(n).

Let $L_1, L_2, L_3, \ldots, L_{n-1}$ be a complete set of MOLS(n). Let $P = \{(i, j) \mid 1 \leq i, j \leq n\}$ and arrange these n^2 ordered pairs in an $n \times n$ grid A as follows:

A =

(1, n)	(2, n)	(3,n)	···	(n,n)
⋮	⋮	⋮		⋮
(1,3)	(2,3)	(3,3)	···	(n,3)
(1,2)	(2,2)	(3,2)	···	(n,2)
(1,1)	(2,1)	(3,1)	···	(n,1)

Figure 6.4: *Naming the cells of A.*

Define a collection of subsets B of P, each consisting of n ordered pairs as follows:

(1) Each of the n columns of A belongs to B.

(2) Each of the n rows of A belongs to B.

(3) For *each* L_i, each of the symbols $1, 2, \ldots, n$ determines a transversal of A. Place each of these n transversals in B.

Then (P, B) is an affine plane of order n. The n lines in (1) are a parallel class, the n lines in (2) are a parallel class, and the n lines determined by each latin square are a parallel class (for a total of $n^2 + n$ lines and $n + 1$ parallel classes).

Example 6.4.1 (Construction of an affine plane from a complete set of MOLS(4)).

1	2	4	3
2	1	3	4
4	3	1	2
3	4	2	1

L_1

1	3	4	2
4	2	1	3
2	4	3	1
3	1	2	4

L_2

1	4	2	3
3	2	4	1
4	1	3	2
2	3	1	4

L_3

A =

14	24	34	44
13	23	33	43
12	22	32	42
11	21	31	41

(1) $\{11, 12, 13, 14\}$ (2) $\{14, 24, 34, 44\}$
$\{21, 22, 23, 24\}$ $\{13, 23, 33, 43\}$
$\{31, 32, 33, 34\}$ $\{12, 22, 32, 42\}$
$\{41, 42, 43, 44\}$ $\{11, 21, 31, 41\}$

L_1 $\{14, 23, 32, 41\}$ L_2 $\{14, 21, 33, 42\}$
$\{13, 24, 31, 42\}$ $\{12, 23, 31, 44\}$
$\{11, 22, 33, 44\}$ $\{11, 24, 32, 43\}$
$\{12, 21, 34, 43\}$ $\{13, 22, 34, 41\}$

L_3 $\{14, 22, 31, 43\}$
$\{11, 23, 34, 42\}$
$\{13, 21, 32, 44\}$
$\{12, 24, 33, 41\}$

Not too surprisingly, we can reverse the above construction to produce a complete set of MOLS(n). To be specific: Let (P, B) be an affine plane of order

n. Label the $n+1$ parallel classes in B by $V, H, \pi_1, \pi_2, \ldots, \pi_{n-1}$ and label the lines in each parallel class with the integers $1, 2, 3, \ldots, n$. For each parallel class π_x we construct a latin square L_x of order n as follows: Fill in cell (i, j) of L_x with the label of the line in π_x which contains the point of intersection of line i in V with line j in H. The resulting collection of latin squares is a complete set of MOLS(n).

Example 6.4.2 **(Construction of a complete set of MOLS(n) from an affine plane.)**

We use the affine plane of order 3 from Example 6.1.1(b) to produce a complete set of MOLS(3). First label the parallel classes (arbitrarily) with V, H, π_1 and π_2. Then, within each parallel class, label the lines (again arbitrarily) with $1, 2$ and 3.

V		H		π_1		π_2	
1 2 3	1	1 4 7	1	1 5 9	1	1 6 8	1
4 5 6	2	2 5 8	2	2 6 7	2	2 4 9	2
7 8 9	3	3 6 9	3	3 4 8	3	3 5 7	3

Now use π_1 to form L_1 and π_2 to form L_2 with the cells named as in Figure 6.4. For example, to fill cell $(3, 2)$ of L_2, we first find the point of intersection of line 3 of V and line 2 of H, namely symbol 8. Since 8 is in line 1 of π_2, cell $(3, 2)$ of L_2 contains 1.

L_1

3	2	1
2	1	3
1	3	2

L_2

3	1	2
2	3	1
1	2	3

The naming of the cells in L_1 and L_2 above is chosen to be from the lower left hand corner (see Figure 6.4), but of course any naming will do nicely. For example, if we name the cells from the upper left hand corner (as we do in all previous chapters) the resulting squares are:

L_1

1	2	3
3	1	2
2	3	1

L_2

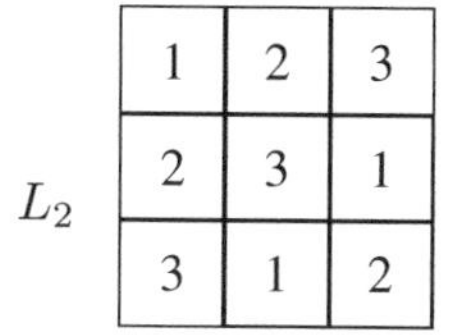

1	2	3
2	3	1
3	1	2

We will see that naming the cells of L_i with cell $(1, 1)$ in the lower left hand corner will be of use in Section 6.5.

Theorem 6.4.3 *An affine plane of order n (and therefore a projective plane of order n) is equivalent to a complete set of MOLS(n).* □

Since there is a complete set of MOLS(n) of every order $n = p^\alpha > 2$ for every *prime p* (constructed from finite fields), there is an *affine plane* (and therefore a *projective* plane) for each of these orders. These are the *ONLY* orders for which affine planes are known to exist. Although a theorem due to R. H. Bruck and H. J. Ryser rules out the existence of affine planes of certain non-prime power orders, there are plenty of unsettled cases remaining (to put it mildly).

Theorem 6.4.4 (R. H. Bruck and H. J. Ryser [4].) *Let $n \equiv 1$ or 2 (mod 4) and let the squarefree part of n contain at least one prime factor $p \equiv 3$ (mod 4). Then there* does not exist *an affine plane of order n.* □

This theorem rules out affine planes of orders $6, 14$, and 22 among others.

Fairly recently $n = 10$ was ruled out (by means of a massive computer search) and so the first unsettled case is $n = 12$. In order to construct an affine plane of order 12, it is necessary to construct 11 MOLS(12). The closest anyone has come is 5 MOLS(12). This is due to A. L. Dulmage, D. M. Johnson, and N. S. Mendelsohn [6] and was done by hand!

Open Problem. Does there exist an affine plane of non-prime power order?

Exercises

6.4.5 Let $L_1, L_2, \ldots, L_{n-1}$ be a complete set of MOLS(n) and let $i \neq p$, $j \neq q$. Prove that the cells (i, j) and (p, q) are occupied by the same symbol in *exactly* one of the latin squares $L_1, L_2, \ldots, L_{n-1}$.

6.4.6 Prove that the pair (P, B) constructed from a complete set of MOLS(n) is in fact an affine plane. (Hint: count the number of blocks and show that every pair of points (actually ordered pairs) belong to at least one of the blocks constructed.)

6.4.7 Prove that the arrays constructed from an affine plane of order n are indeed a complete set of MOLS(n).

6.4.8 Construct 3 MOLS(4) from the affine plane of order 4 in Exercise 6.3.4.

6.4.9 Use the complete set of MOLS in Example 5.2.4 and Exercises 5.2.6, 5.2.7, and 5.2.8 to construct affine planes.

6.5 Coordinatizing the affine plane.

Let (P, B) be an affine plane of order n. Then, of course, $|P| = n^2$, $|B| = n$, and $|B| = n^2 + n$. Label the parallel classes $V, H, \pi_1, \pi_2, \ldots, \pi_{n-1}$ and label the lines in V and H:

$$\begin{cases} V: & x = 0, x = 1, x = 2, \ldots, x = n = 1 \\ H: & y = 0, y = 1, y = 2, \ldots, y = n - 1. \end{cases}$$

Then we say the point $p \in P$ has coordinates (i, j) if and only if p belongs to the line $x = i$ *and* the line $y = j$

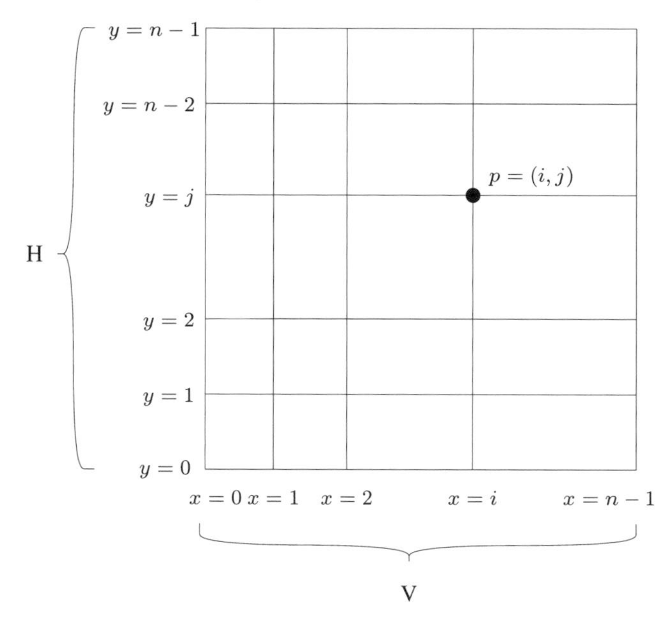

Call the line $y = 0$ the x-*axis*, the line $x = 0$ the y-axis, the line $x = 1$ the *line of slopes*, and the point $(0, 0)$ the *origin*.

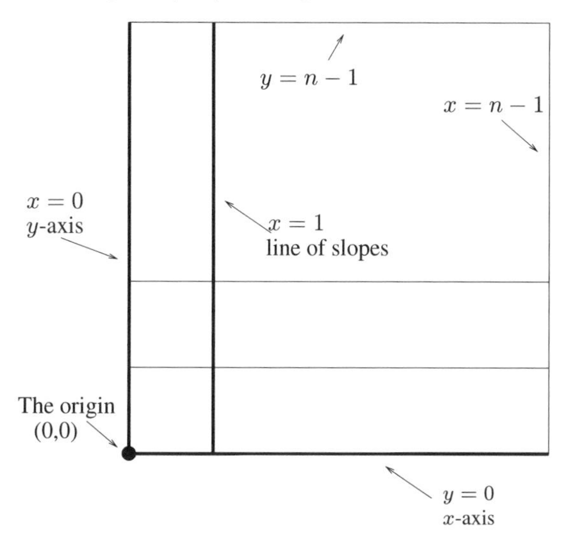

Each parallel class π_i has *exactly* one line ℓ containing the *origin*. Since $\ell \notin V$ or H, ℓ must *intersect* the line of slopes $x = 1$. Let $(1, m)$ be the point of intersection of ℓ and the line of slopes $x = 1$.

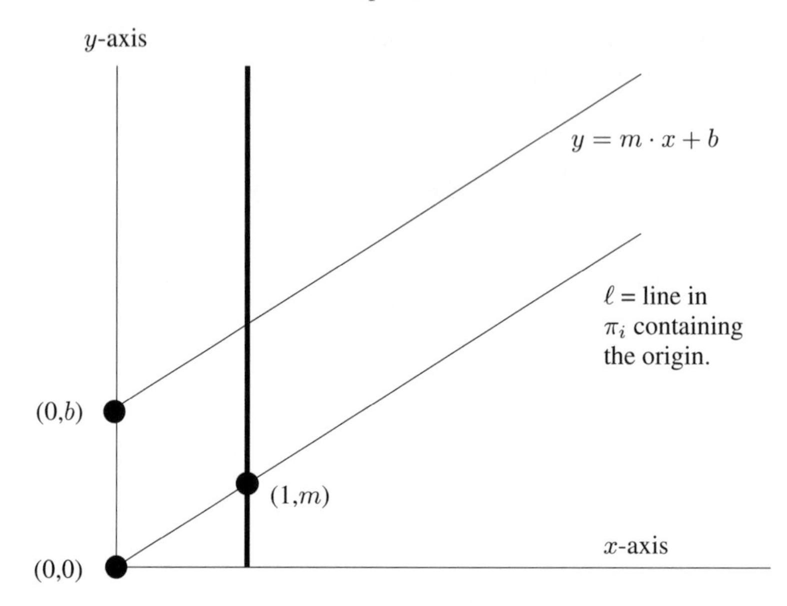

We will say that *each* line in the parallel class π_i has *slope* m. Additionally we will say that each line in V has no slope *and* each line in H has slope 0. We can now label each line k in the parallel class π_i with an equation of the form $y = m \cdot x + b$, where m is the slope of the parallel class π_i and $(0, b)$ is the point of intersection of k with the y-axis. (Since the y-axis belongs to the parallel class V, every line in π_i intersects the y-axis.) Just as in ordinary analytic geometry, the equation $y = m \cdot x + b$ is called the *point slope* formula.

Example 6.5.1 Let (P, B) be the affine plane of order 3 given below.

147	123	159	168
258	456	267	249
369	789	348	357
V	H	π_1	π_2

$\{1, 4, 7\}$	x = 0	y-axis	$\{1, 2, 3\}$	y = 0	x-axis
$\{2, 5, 8\}$	x = 1	line of slopes	$\{4, 5, 6\}$	y = 1	
$\{3, 6, 9\}$	x = 2		$\{7, 8, 9\}$	y = 2	

Then $1 = (0, 0)$ origin, $2 = (1, 0), 3 = (2, 0), 4 = (0, 1), 5 = (1, 1), 6 = (2, 1)$, $7 = (0, 2), 8 = (1, 2)$, and $9 = (2, 2)$.

To find the point slope formula for the line $k = \{3, 4, 8\} \in \pi_1$, we first find the line in π_1 containing the origin $(0, 0) = 1$. This is $\ell = \{1, 5, 9\}$. The intersection of ℓ with the line of slopes $\{2,5,8\}$ is $5 = (1,1)$ and so the slope of every line in π_1 has point slope formula of the form $y = 1 \cdot x + b$. To determine b we must find the intersection of $k = \{3,4,8\}$ with the y-axis $\{1,4,7\}$. This is $4 = (0,1)$ and so $b = 1$ and the point slope formula for $k = \{3,4,8\}$ is $y = 1 \cdot x + 1$. Similarly, the point slope formulas for the other lines in π_1 and π_2 are:

$\{1,5,9\}$	$y = 1 \cdot x + 0$	$\{1,6,8\}$	$y = 2 \cdot x + 0$
$\{2,6,7\}$	$y = 1 \cdot x + 2$	$\{2,4,9\}$	$y = 2 \cdot x + 1$
$\{3,4,8\}$	$y = 1 \cdot x + 1$	$\{3,5,7\}$	$y = 2 \cdot x + 2$

Exercises

6.5.2 Coordinatize each of the affine planes constructed in Exercise 6.4.8 and compute the point slope formula for each of the lines *not* in V and H.

Chapter 7

Steiner Quadruple Systems

7.1 Introduction

A *Steiner Quadruple System* is an ordered pair (V, B) where V is a finite set of *symbols* and B is a set of 4-element subsets of V called *quadruples* with the property that every 3-element subset of V is a subset of exactly one quadruple in B. $|V|$ is called the *order* of the Steiner Quadruple System. For short, we write $SQS(v)$ to denote a Steiner Quadruple System of order v.

It is very important to remember that two quadruples in a Steiner Quadruple System *can* intersect in two symbols, since it is only the 3-element subsets that are required to occur in a unique quadruple. (For those of us who are most familiar with block designs, this is easy to forget!)

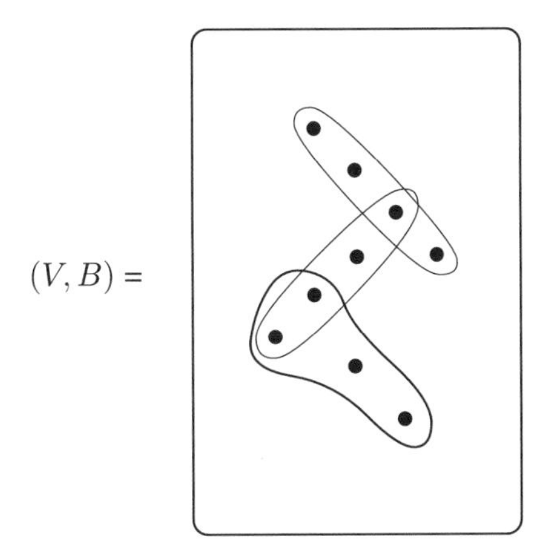

Figure 7.1: *Quadruples in a Steiner Quadruple System.*

In this chapter we will find the set of integers v for which there exists a Steiner quadruple system of order v.

We can easily find the number of quadruples in any Steiner quadruple system (V, B) of order v. Each quadruple $\{a, b, c, d\}$ in B contains the four 3-element subsets $\{a, b, c\}$, $\{a, b, d\}$, $\{a, c, d\}$ and $\{b, c, d\}$, and altogether there are $\binom{v}{3}$ 3-element subsets of V. Since the blocks in B partition these $\binom{v}{3}$ 3-element subsets, we have the following result.

Theorem 7.1.1 *The number of quadruples in a $SQS(v)$, (V, B) is*

$$|B| = \binom{v}{3}/4 = v(v-1)(v-2)/24.$$

As in the case of Steiner triple systems, knowing $|B|$ helps in proving that (V, B) is a $SQS(v)$: if each 3-element subset of V occurs in *at least* one block, *and* if $|B| = v(v-1)(v-2)/24$, then (V, B) is a SQS (see Exercise 7.1.9).

Example 7.1.2 (Trivial examples.)

(a) $(V, B) = (\{1\}, \emptyset)$ is a $SQS(1)$.

(b) $(V, B) = (\{1, 2\}, \emptyset)$ is a $SQS(2)$, because V contains no 3-element subsets.

(c) $(V, B) = (\{1, 2, 3, 4\}, \{\{1, 2, 3, 4\}\})$ is a $SQS(4)$.

The Boolean Steiner Quadruple System Construction (for a $SQS(2^n)$) For each $n \geq 3$, there exists a $SQS(2^n)$ (V, B) which can be constructed as follows. Let $V = Z_2^n$ be the set of binary words of length n (that is, vectors of length n, each component of which is 0 or 1). If $a = (a_1, \ldots, a_n)$ and $b = (b_1, \ldots, b_n)$ are two binary words of length n, then we define $a + b = (a_1 + b_1, \ldots, a_n + b_n)$, reducing each sum modulo 2. So $a + b$ is another binary word of length n. (For example, $(0, 1, 1, 0) + (0, 0, 1, 1) = (0, 1, 0, 1)$; for brevity we write $0110 + 0011 = 0101$.) Now define

$$B = \{\{a, b, c, d\} | a, b, c \text{ and } d \text{ are 4 distinct binary words of length } n \text{ such that } a + b + c + d = 00\ldots0\}.$$

Then (V, B) is the Steiner quadruple system known as the *Boolean Steiner quadruple system.*

Proof To see that this does define a $SQS(v = 2^n)$, suppose that a, b and c are any three different binary words of length n (i.e., three elements of V). Since each quadruple in B has the property that the sum of the 4 elements is $00\ldots0$, the only possible 4-element subset of V that could be in B and contain $\{a, b, c\}$ is $\{a, b, c, d\}$ where $d = a + b + c$ (so therefore $a + b + c + d = 00\ldots0$). So $\{a, b, c\}$ is in at most one quadruple in B. But to be in B, we also need to show that a, b, c and d are all different. Clearly this is so, since we chose a, b and c to all be different, and if d is one of the others, say $d = a$, then $a+b+c+d = 00\ldots0$, which implies that $c = b$; so indeed a, b, c and d are all different. Therefore $\{a, b, c\}$ is in exactly one quadruple in B, so (V, B) is a $SQS(2^n)$. □

Example 7.1.3 We can find the Boolean Steiner quadruple system (V, B) of order 8 as follows. Since $8 = 2^3$, V consists of all binary words of length 3; so

$$V = \{000, 001, 010, 011, 100, 101, 110, 111\}.$$

We find the $|B| = 8 \cdot 7 \cdot 6/24 = 14$ quadruples by making sure that if a, b, c and d are all different, and if $a + b + c + d = 0$, then $\{a, b, c, d\}$ is a block. So for example, if $a = 100$, $b = 101$, $c = 110$ and $d = 111$, then they are all different, and $a+b+c+d = 100+101+110+111 = 000$; so $\{100, 101, 110, 111\} \in B$. Thus we obtain the following blocks.

$$\begin{array}{ll}
\{000, 001, 010, 011\} & \{100, 101, 110, 111\} \\
\{000, 001, 100, 101\} & \{010, 011, 110, 111\} \\
\{000, 001, 110, 111\} & \{010, 011, 100, 101\} \\
\{000, 010, 100, 110\} & \{001, 011, 101, 111\} \\
\{000, 010, 101, 111\} & \{001, 011, 100, 110\} \\
\{000, 011, 100, 111\} & \{001, 010, 101, 110\} \\
\{000, 011, 101, 110\} & \{001, 010, 100, 111\}
\end{array}$$

Example 7.1.4 Construct a $SQS(10)$, (V, B) as follows. Let V be the set of edges of K_5, the complete graph on 5 vertices; so $V = \{\{1,2\}, \{1,3\}, \{1,4\}, \{1,5\}, \{2,3\}, \{2,4\}, \{2,5\}, \{3,4\}, \{3,5\}, \{4,5\}\}$. Let B consist of the sets of 4 edges, each of which form a graph that is isomorphic to one of the following graphs: the complete bipartite graph $K_{1,4}$; the 4-cycle C_4; or $K_2 + K_3$ (the vertex-disjoint union of K_2 and K_3). These graphs can be represented graphically as follows.

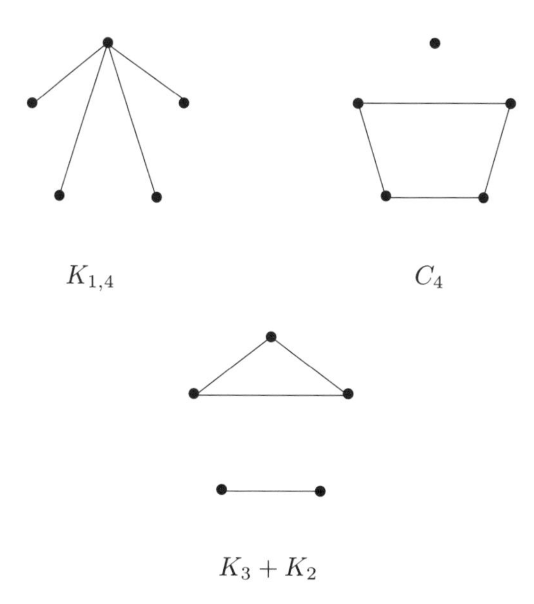

So, for example, we obtain the following quadruples and their corresponding labelled graphs

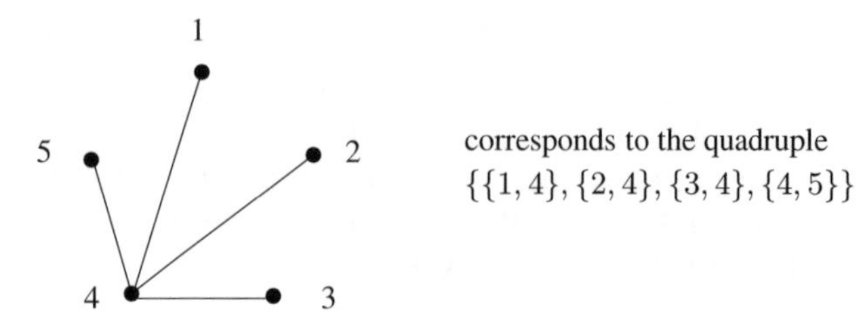

corresponds to the quadruple
$\{\{1,4\},\{2,4\},\{3,4\},\{4,5\}\}$

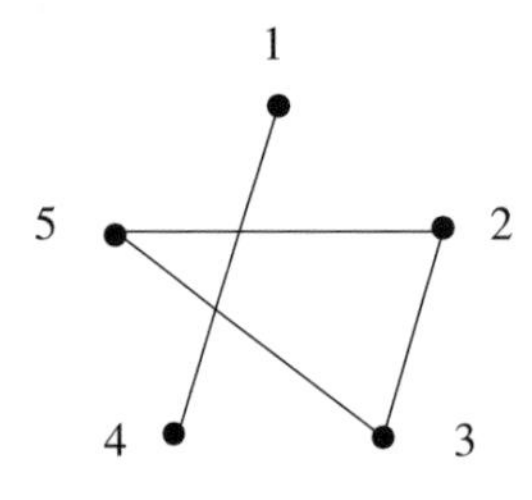

corresponds to the quadruple
$\{\{1,4\},\{2,3\},\{2,5\},\{3,5\}\}$

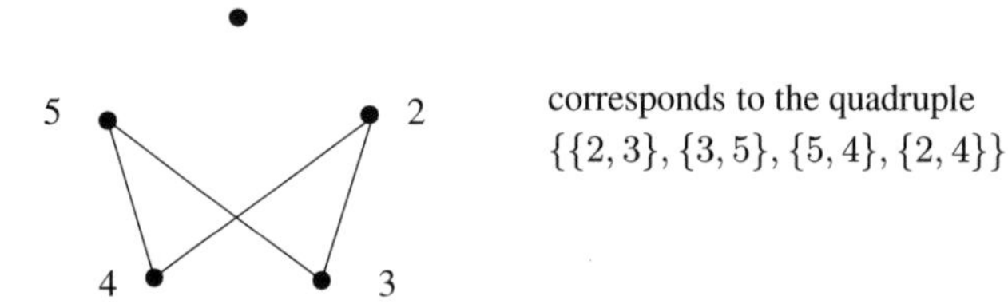

corresponds to the quadruple
$\{\{2,3\},\{3,5\},\{5,4\},\{2,4\}\}$

Showing that this defines a $SQS(10)$ is left to Exercise 7.1.6.

Example 7.1.5 We construct a $SQS(14)$, (V, B) as follows. Let $V = \{1,2,3,4,5,6,7\} \times \{1,2\}$. Define

$$\begin{aligned}
b_1 &= \{(1,1),(4,1),(1,2),(4,2)\} \\
b_2 &= \{(1,1),(4,1),(2,2),(6,2)\} \\
b_3 &= \{(1,1),(4,1),(3,2),(7,2)\} \\
b_4 &= \{(1,1),(3,1),(4,1),(5,2)\} \\
b_5 &= \{(1,1),(4,1),(5,1),(6,1)\}
\end{aligned}$$

Also define permutations $\pi_i : V \to V$ for $1 \le i \le 3$ by

$$\begin{aligned}\pi_1((x,a)) &= (x+1,a),\\ \pi_2((x,a)) &= (2x,a), \text{ and}\\ \pi_3((x,a)) &= (-x,a+1),\end{aligned}$$

reducing the first component modulo 7 and the second component modulo 2. For each quadruple $b = \{w,x,y,z\}$, define $\pi_i(b) = \{\pi_i(w),\pi_i(x),\pi_i(y),\pi_i(z)\}$. Then clearly $\pi_1^7(b) = b$, $\pi_2^3(b) = b$ and $\pi_3^2(b) = b$. Finally, for $1 \le j \le 5$, define the set of quadruples B_j to be the smallest set satisfying

(a) $b_j \in B_j$, and

(b) for $1 \le i \le 3$, if $b \in B_j$ then $\pi_i(b) \in B_j$.

Define $B = \bigcup_{j=1}^{5} B_j$. Then (V,B) is a $SQS(14)$ (see Exercise 7.1.13).

So, for example, we construct B_1 as follows.

By (a), $b_1 \in B_1$. Next we apply (b) with $i = 1, 2$ and 3, as follows. By (b) with $i = 1$, since $b_1 \in B_1$, so are $\pi_1(b_1), \pi_1^2(b_1), \ldots, \pi_1^6(b_1)$; so B_1 contains the seven quadruples $\{(1,1),(4,1),(1,2),(4,2)\}, \{(2,1),(5,1),(2,2),(5,2)\}, \ldots, \{(7,1),(3,1),(7,2),(3,2)\}$ respectively.

Then, by (b) with $i = 2$, B_1 contains $\pi_2(b_1) = \{(2,1),(1,1),(2,2),(1,2)\}$ and $\pi_2^2(b_1) = \{(4,1),(2,1),(4,2),(2,2)\}$, and 12 more quadruples formed by applying $\pi_1, \pi_1^2, \ldots, \pi_1^6$ to each of these two quadruples, $\pi_2(b_1)$ and $\pi_2(b_2)$. Finally, by (b) with $i = 3$, $\pi_3(b_1) \in B_1$. But we notice that

$$\pi_3(b_1) = \{(6,2),(3,2),(6,1),(3,1)\} = \pi_1^2(b_1)!$$

So B_1 contains just the 21 quadruples defined so far (you should check that for any of these 21 quadruples b, $\pi_1(b) \in B_1$, $\pi_2(b) \in B_1$, and $\pi_3(b) \in B_1$). Similarly one can check that (see Exercise 7.1.12) $|B_1| = |B_2| = |B_3| = 21$ and $|B_4| = |B_5| = 14$.

Exercises

7.1.6 Find all the blocks of the $SQS(10)$ in Example 7.1.4, and verify that an $SQS(10)$ is indeed produced.

7.1.7 Check that the Boolean Steiner quadruple system defined in Example 7.1.2 is indeed a Steiner quadruple system.

7.1.8 Show that if x is a symbol in a $SQS(v)$, then the number of quadruples that contain x is $(v-1)(v-2)/6$.

7.1.9 Show that if each 3-element subset of (V,B) is contained in *at least* one block of B, and if the number of blocks in (V,B) is $v(v-1)(v-2)/24$, then (V,B) is a SQS.

7.1.10 In the Boolean $SQS(8)$ in Example 7.1.3, find all the blocks that contain 000, delete 000 from these blocks, and show that what remains forms a $STS(7)$.

7.1.11 Show that it is possible to rename the symbols in the $SQS(10)$ in Example 7.1.4 with $\{1, 2, \ldots, 10\}$ in such a way that, after renaming the symbols, B contains the quadruples $\{1, 2, 3, 10\}, \{4, 5, 6, 10\}$, $\{7, 8, 9, 10\}$ and $\{3, 6, 9, 10\}$. (This property will be used in The $(3v - 2)$ Construction in Section 7.2.)

7.1.12 Show that in the $SQS(14)$ constructed in Example 7.1.5, $|B_1| = |B_2| = |B_3| = 21$ and $|B_4| = |B_5| = 14$. Notice that $\sum_{j=1}^{5} |B_j| = 91 = (14)(13)(12)/24$ as required by Exercise 7.1.9.

7.1.13 Show that the set of quadruples B defined in Example 7.1.5 do form a $SQS(14)$. (Hint: To do this, first show that each of the following triples is in a quadruple in B: $\{(1,1),(2,1),(3,1)\}$, $\{(1,1),(2,1),(4,1)\}$, $\{(1,1),(2,1),(5,1)\}$; and therefore using π_2 and π_1 shows that $\{(1,1),(2,1),(6,1)\}$ and $\{(1,1),(3,1),(5,1)\}$ are in quadruples in B. Apply π_1 to each of these quadruples to see that each of the 35 triples containing symbols with second coordinate 1 occurs in a quadruple in B. Applying π_3 to each of these quadruples shows that each of the 35 triples containing symbols with second coordinate 2 occurs in a quadruple in B. Finally, show that for $1 \le i \le 7$, B contains quadruples containing the triples $\{(1,1),(2,1),(i,2)\}$. Therefore, by using π_2 and π_1, show that $\{(1,1),(3,1),(i,2)\}$ and $\{(1,1),(4,1),(i,2)\}$ are in quadruples in B, and so by using π_1 and π_3 show that $\{(1,2),(2,2),(i,1)\}$, $\{(1,2),(3,2),(i,1)\}$, and $\{(1,2),(4,2),(i,1)\}$ are in quadruples in B. Then applying π_1 to each of these quadruples shows that each of the remaining $(7)(7)(6) = 294$ triples occurs in a quadruple in B.)

We now begin by finding some necessary conditions for the existence of a $SQS(v)$, and then proceed to show that these conditions are sufficient by producing several recursive constructions.

Lemma 7.1.14 *If there exists a $SQS(v)$, then either $v = 1$ or $v \equiv 2$ or 4 (mod 6).*

Remark What we prove here actually shows that if (V, B) is a $SQS(n)$ with $n > 1$ then for any $v \in V$, the blocks in B that contain v form a $STS(n-1)$ when v is deleted from them.

Proof Let (V, B) be a $SQS(v)$. If $B = \emptyset$ then $v \le 2$. So we can assume that $B \neq \emptyset$ and $v \ge 4$.

Let $p \in V$ and define $B(p)$ to be the set of triples formed by taking all the quadruples in B that contain p, then deleting p; that is

$$B(p) = \{b' \backslash \{p\} | b' \in B, p \in b'\}.$$

Clearly if $b \in B(p)$ then $|b| = 3$, so $B(p)$ is a collection of triples of V. (For example, see Exercise 7.1.10.) In fact, $(V \backslash \{p\}, B(p))$ is a STS as we will now see.

Let x and y be two symbols in $V\backslash\{p\}$. Since (V, B) is a $SQS(v)$, $\{p, x, y\}$ occurs in exactly one block $b' \in B$, and so $\{x, y\}$ occurs in exactly one block of $B(p)$. Therefore $(V\backslash\{p\}, B(p))$ is a $STS(v-1)$, which only exists if $v-1 \equiv 1$ or 3 (mod 6) (see Theorem 1.1.3). Therefore $v \equiv 2$ or 4 (mod 6). □

Remark The Steiner triple system formed in the proof of Lemma 7.1.14 is called a *derived* STS of the SQS. A famous unsolved conjecture is to show that every Steiner triple system is a derived $STS(v)$ of some $SQS(v+1)$. This has been shown to be true whenever $v \leq 15$ [5].

The first proof that for all $v \equiv 2$ or 4 (mod 6) there exists a $SQS(v)$ was given by Hanani [8] in 1960. The proof we give here uses some of his constructions, but is the proof of Lenz [15]. A little bit of graph theory is used in these constructions so we introduce it when necessary. We end this section with some graph theory.

A 1-*factor* of a graph G is a set of edges F in G with the property that each vertex is incident with exactly one edge in F. A 1-*factorization* of a graph G is a partition of the edge set of G into 1-factors.

Lemma 7.1.15 *K_{2x} has a 1-factorization.*

Proof Let K_{2x} be defined on the vertex set $\{\infty\} \cup \{1, 2, 3, \ldots, 2x-1\}$. For $1 \leq i \leq 2x-1$ define

$$F_i = \{\{\infty, i\}, \{-1+i, 1+i\}, \{-2+i, 2+i\}, \ldots, \{-(x-1)+i, (x-1)+i\}\},$$

reducing each term modulo $2x-1$. Then $\{F_1, F_2, \ldots, F_{2x-1}\}$ is a 1-factorization of K_{2x}. □

For later use, a different 1-factorization will be useful. The following 1-factorization can be constructed by using Steiner triple systems.

Lemma 7.1.16 *For all $v \equiv 2$ or 4 (mod 6) with $v \geq 4$, there exists a 1-factorization $\{F_2, \ldots, F_v\}$ of K_v in which $\{\{1, 2\}, \{3, 4\}\} \subseteq F_2$, $\{\{1, 3\}, \{2, 4\}\} \subseteq F_3$ and $\{\{1, 4\}, \{2, 3\}\} \subseteq F_4$, where K_v is defined on the vertex set $\{1, 2, \ldots, v\}$.*

Remark Such 1-factorizations of K_v exist for all even $v \geq 8$, but this result will suffice for our purposes.

Proof Let $(\{2, 3, \ldots, v\}, T)$ be a $STS(v-1)$. By renaming the symbols if necessary, we can assume that $\{2, 3, 4\} \in T$. For $2 \leq j \leq v$ define

$$F_j = \{\{1, j\}\} \bigcup \{\{a, b\} | \{j, a, b\} \in T\}.$$

Then obviously the edge $\{1, j\}$ occurs in F_j. Each pair $\{a, b\} \subseteq \{2, 3, \ldots, v\}$ occurs in a triple, say $\{a, b, j\}$ in the $STS(v-1)$, so then the edge $\{a, b\}$ occurs in F_j. Therefore each edge of K_v occurs in exactly one of $F_2, F_3, \ldots, F_v$, so these sets do partition the edges of K_v.

It remains to see that F_j is a 1-factor of K_v for $2 \leq j \leq v$, so let a be any vertex of K_v. If $a = 1$ or j then F_j contains the edge $\{1, j\}$. Otherwise $a \in \{2, 3, \ldots, v\} \backslash j$. In this case, the pair $\{a, j\}$ occurs in some triple, say $\{a, j, b\}$ of the $STS(v-1)$, so F_j contains the edge $\{a, b\}$. So each vertex is incident with exactly one edge in F_j, and therefore F_j is a 1-factor.

Therefore $\{F_2, \ldots, F_v\}$ is a 1-factorization of K_v. Clearly F_2, F_3 and F_4 contain the edges required in the lemma, since $\{2, 3, 4\}$ is a triple in the $STS(v-1)$. So the result is proved. □

Example 7.1.17 We can use Lemma 7.1.16 to construct a 1-factorization of K_8. A $STS(7)$ defined on the vertex set $\{2, 3, \ldots, 8\}$ in which T contains $\{2, 3, 4\}$ can be defined as follows:

$$T = \{\{2,3,4\}, \{3,5,6\}, \{4,5,7\}, \{4,6,8\}, \{2,6,7\}, \{3,7,8\}, \{2,5,8\}\}.$$

(Simply rename symbols in your favorite $STS(7)$.) Then F_j is formed from the triples containing symbol j as defined in the proof of Lemma 7.1.16. So for example, to find $F_5 = F_j$, we first find all the triples in T containing symbol $j = 5$: $\{3,5,6\}$, $\{4,5,7\}$, and $\{2,5,8\}$. Deleting symbol $j = 5$ from each of these triples then adding the edge $\{i, j\} = \{1, 5\}$ produces the 1-factor F_4:

$$F_5 = \{\{1,5\}, \{3,6\}, \{4,7\}, \{2,8\}\}.$$

Similarly we can find

$$\begin{aligned} F_2 &= \{\{1,2\}, \{3,4\}, \{6,7\}, \{5,8\}\}, \\ F_3 &= \{\{1,3\}, \{2,4\}, \{5,6\}, \{7,8\}\}, \text{ and} \\ F_4 &= \{\{1,4\}, \{2,3\}, \{5,7\}, \{6,8\}, \end{aligned}$$

and these contain the edges described in Lemma 7.1.16. (Exercise 7.1.18 asks you to find the remaining 1-factors.)

Exercises

7.1.18 Find the remaining 1-factors F_6, F_7 and F_8 in the 1-factorization of K_8 defined in Example 7.1.17.

7.1.19 Use the proof of Lemma 7.1.15 to find a 1-factorization of K_6.

7.1.20 Use the proof of Lemma 7.1.16 to find a 1-factorization of K_{10}.

7.2 Constructions of Steiner Quadruple Systems

We are now ready to present the first recursive construction of Steiner quadruple systems.

The $2v$ Construction (to produce a $SQS(2v)$ from a $SQS(v)$)
Let (V, B) be a $SQS(v)$. Let $\{F_1, \ldots, F_{v-1}\}$ be a 1-factorization of K_v defined on the vertex set V. Define a $SQS(2v)$ $(V \times \{1, 2\}, B')$ as follows:

Type 1: for each $b \in B$ and each $i \in \{1, 2\}$, let $b \times \{i\}$ be in B', and

Type 2: for $1 \leq j \leq v - 1$, for each edge $\{a, b\} \in F_j$ and for each edge $\{c, d\} \in F_j$ (including $\{a, b\} = \{c, d\}$), let $\{(a, 1), (b, 1), (c, 2), (d, 2)\}$ be in B'.

For a pictorial representation of this construction, see Figure 7.2.

Theorem 7.2.1 *The $2v$ Construction produces a $SQS(2v)$. Moreover, if $v \geq 4$ then $\{F_1, \ldots, F_{v-1}\}$ can be chosen so that the $SQS(2v)$ contains a $SQS(8)$.*

Proof To see that $(V \times \{1, 2\}, B')$ defined in The $2v$ Construction does define a 3-design, we show that each 3-element subset of $V \times \{1, 2\}$ is in at least one quadruple in B', and then show that B' contains $2v(2v - 1)(2v - 2)/24$ quadruples (see Exercise 7.1.9).

Suppose that all three of the symbols in the 3-element subset have the same second coordinate; say the symbols are (x, i), (y, i) and (z, i). Then since (V, B) is a SQS, $\{x, y, z\}$ is contained in a quadruple $\{w, x, y, z\} \in B$, so from (1) we have that $\{(w, i), (x, i), (y, i), (z, i)\} \in B'$.

Otherwise, two of the second coordinates of the 3 symbols must be the same, so we may assume that the 3-element subset under consideration is $\{(x, 1), (y, 1), (z, 2)\}$ or $\{(x, 2), (y, 2), (z, 1)\}$. The edge $\{x, y\}$ occurs in exactly one 1-factor, say F_j in the 1-factorization of K_v. Since F_j is a 1-factor, it also contains exactly one edge incident with vertex z, say $\{w, z\}$. Then from (2), $\{(x, 1), (y, 1), (z, 2), (w, 2)\} \in B'$ and $\{(x, 2), (y, 2), (z, 1)\} \in B'$.

Therefore each 3-element subset of $V \times \{1, 2\}$ occurs in at least one quadruple in B'. By Exercise 7.2.3, B' contains exactly $2v(2v-1)(2v-2)/24$ quadruples. So by Exercise 7.1.9, $(V \times \{1, 2\}, B')$ is a $SQS(2v)$ as required.

We can rename the symbols in V so that $V = \{1, 2, \ldots, v\}$ and so that $\{1, 2, 3, 4\} \in B$. If we construct $\{F_1, \ldots, F_{v-1}\}$ using Lemma 7.1.16, then since $\{\{1, 2\}, \{3, 4\}\} \subseteq F_1, \{\{1, 3\}, \{2, 4\}\} \subseteq F_2$ and $\{\{1, 4\}, \{2, 3\}\} \subseteq F_3$, and since $\{1, 2, 3, 4\} \in B$, it is clear that The $2v$ Construction contains a $SQS(8)$ on the symbols $\{1, 2, 3, 4\} \times \{1, 2\}$. □

(1) Type 1 quadruples.

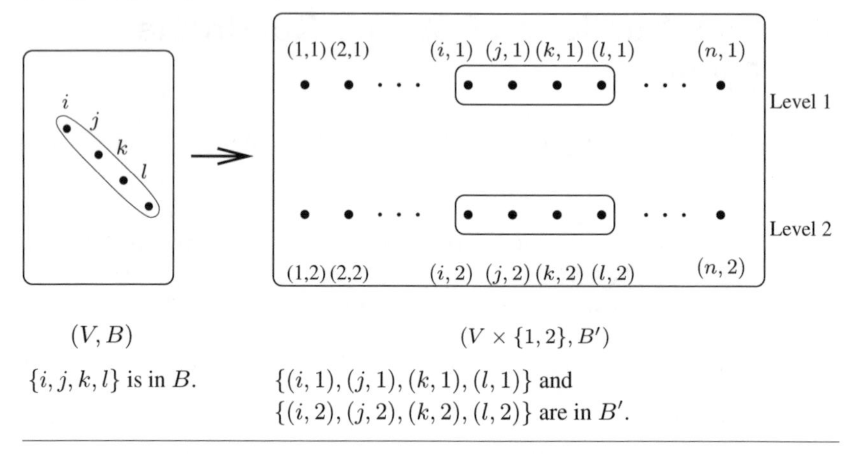

(2) Type 2 quadruples.

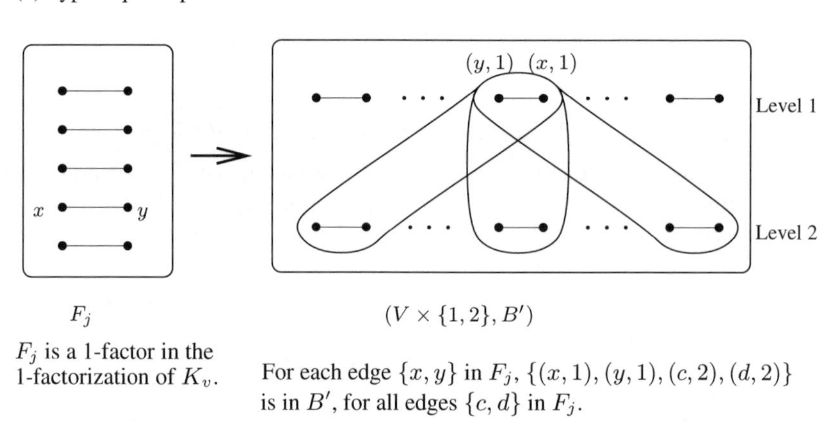

Figure 7.2: *The $2v$ Construction.*

Example 7.2.2 $(\{1,2,3,4\}, \{\{1,2,3,4\}\})$ is a $SQS(4)$. $F_1 = \{\{1,2\}, \{3,4\}\}$, $F_2 = \{\{1,3\},\{2,4\}\}$ and $F_3 = \{\{1,4\},\{2,3\}\}$ form a 1-factorization of K_4. Using The $2v$ Construction, we produce a $SQS(8)(\{1,2,3,4\} \times \{1,2\}, B')$ as follows:

(1) $\{(1,1),(2,1),(3,1),(4,1)\}$ and $\{(1,2),(2,2),(3,2),(4,2)\}$ are in B', and

(2) B' contains the following quadruples

$$\begin{array}{cc}
j=1 & j=2 \\
\{(1,1),(2,1),(1,2),(2,2)\} & \{(1,1),(3,1),(1,2),(3,2)\} \\
\{(1,1),(2,1),(3,2),(4,2)\} & \{(1,1),(3,1),(2,2),(4,2)\} \\
\{(3,1),(4,1),(1,2),(2,2)\} & \{(2,1),(4,1),(1,2),(3,2)\} \\
\{(3,1),(4,1),(3,2),(4,2)\} & \{(2,1),(4,1),(2,2),(4,2)\}
\end{array}$$

$$\begin{array}{c}
j=3 \\
\{(1,1),(4,1),(1,2),(4,2)\} \\
\{(1,1),(4,1),(2,2),(3,2)\} \\
\{(2,1),(3,1),(1,2),(4,2)\} \\
\{(2,1),(3,1),(2,2),(3,2)\}.
\end{array}$$

Exercises

7.2.3 Show that the $2v$ Construction defines exactly $2v(2v-1)(2v-2)/24$ quadruples.

7.2.4 Let $(\{1,2,\ldots,8\},B)$ be the $SQS(8)$ formed by renaming the symbols $(1,1),(2,1),(3,1),(4,1),(1,2)$ $(2,2),(3,2)$ and $(4,2)$ in Example 7.2.2 with $1,2,3,4,5,6,7$ and 8 respectively. Let $\{F_1,\ldots,F_7\}$ be the 1-factorization of K_8 defined in Example 7.1.17. This $SQS(8)$ and this 1-factorization of K_8 are used in the $2v$ Construction to produce a $SQS(16)$, (V',B'). Find the quadruple containing the following three symbols in V'.

(a) (1, 1), (4, 1) and (6, 1)

(b) (2, 2), (4, 2) and (5, 2)

(c) (1, 1), (6, 1) and (7, 2)

(d) (3, 1), (2, 2) and (6, 2)

(e) (5, 1), (6, 2) and (7, 2)

Verify that this $SQS(16)$ contains a $SQS(8)$ on the symbols $\{1,2,3,4\}\times\{1,2\}$.

7.2.5 Show that the $2v$ Construction produces a $SQS(2v)$ that contains a $SQS(v)$.

The following is one of Hanani's constructions for a SQS. It is depicted in Figure 7.3.

The $(3v-2)$ Construction (to produce a $SQS(3v-2)$ from a $SQS(v)$) Let $(\{1,2,\ldots,v\},B)$ be a $SQS(v)$. Define a $SQS(3v-2)$ $(V'=\{\infty\}\cup(\{1,2,\ldots,v-1\}\times\{1,2,3\}),B')$ as follows:

Type 1: if $b = \{x, y, z, v\}$ is a quadruple in B containing symbol v, then let $(\{\infty\} \cup (\{x, y, z\} \times \{1, 2, 3\}), B(b))$ be a $SQS(10)$ that contains the quadruples $\{\infty, (x, 1), (x, 2), (x, 3)\}, \{\infty, (y, 1), (y, 2), (y, 3)\}, \{\infty, (z, 1), (z, 2), (z, 3)\}$ and $\{\infty, (x, 3), (y, 3), (z, 3)\}$ (see Exercise 7.1.11) and let $B(b) \setminus \{\{\infty, (i, 1), (i, 2), (i, 3)\} \mid i \in \{x, y, z\}\} \subseteq B'$,

Type 2: if $b = \{w, x, y, z\}$ is a quadruple in B that does not contain v, then let $\{(w, i), (x, j), (y, k), (z, l)\} \in B'$ for all $i, j, k, l \in \{1, 2, 3\}$ such that $i + j + k + l \equiv 0 \pmod 3$, and

Type 3: for $1 \leq x \leq v - 1$, let $\{\infty, (x, 1), (x, 2), (x, 3)\} \in B'$.

Theorem 7.2.6 *The* $(3v - 2)$ *Construction produces a* $SQS(3v - 2)$*. Furthermore, the* $SQS(3v - 2)$ *contains a* $SQS(v)$*.*

Proof We consider the possible 3-element subsets s of $\{\infty\} \cup (\{1, 2, \ldots, v - 1\} \times \{1, 2, 3\})$ in turn.

Suppose $s = \{\infty, (x, i), (x, j)\}$. Then s is in the Type 3 quadruple $\{\infty, (x, 1), (x, 2), (x, 3)\}$.

Suppose $s = \{(x, 1), (x, 2), (x, 3)\}$. Then s is in a Type 3 quadruple.

Suppose $s = \{\infty, (x, i), (y, j)\}$ with $x \neq y$ (possibly $i = j$). Let $\{x, y, a, v\}$ be the unique quadruple in B that contains x, y and v. Then B' contains the Type 1 quadruples in a $SQS(10)$ defined on the set of symbols $\{\infty\} \cup \{\{x, y, a\} \times \{1, 2, 3\}\}$, one of which contains s.

Suppose $s = \{(x, i), (y, j), (z, k)\}$ where exactly two of x, y and z are the same, say $y = z$ (i, j and k do not all need to be different). Again, let $\{x, y, a, v\}$ be the unique quadruple in B containing $\{x, y, v\}$. Then, as in the previous case, B' contains the quadruples in a $SQS(10)$ defined on the set of symbols $\{\infty\} \cup (\{x, y, a\} \times \{1, 2, 3\})$, one of which is $s = \{(x, i), (y, j), (y, k)\}$.

Suppose $s = \{(x, i), (y, j), (z, k)\}$, where x, y and z are all different. Let $\{x, y, z, a\}$ be the unique quadruple in B containing $\{x, y, z\}$. If $a = v$ then s is in a Type 1 quadruple. Otherwise, let $l \in \{1, 2, 3\}$ be chosen so that $i+j+k+l \equiv 0 \pmod 3$. Then $s \subseteq \{(x, i), (y, j), (z, k), (a, l)\}$ which is a Type 2 quadruple.

Therefore each 3-element subset occurs in at least one quadruple, so checking that the number of quadruples defined is $(3v - 2)(3v - 3)(3v - 4)/24$ (see Exercise 7.2.8) completes the proof that a $SQS(3v - 2)$ has been defined (by Exercise 7.1.9).

Making sure that the Type 1 quadruples include $\{\infty, (x, 3), (y, 3), (z, 3)\}$ ensures that the $SQS(3v-2)$ contains a $SQS(v)$ defined on the symbols $(\{\infty\} \cup (\{1, 2, \ldots, v - 1\} \times \{3\})$. To see this, consider any triple $s = \{(x, 3), (y, 3), (z, 3)\}$. We simply need to show that the fourth symbol in the quadruple in B' that contains s is ∞, or has its second coordinate equal to 3. If $\{x, y, z\}$ is in a quadruple of B containing v, then this naming procedure of the Type 1 quadruples ensures that s is in the quadruple $\{\infty, (x, 3), (y, 3), (z, 3)\}$ in B'. Otherwise, $\{x, y, z\}$ is in a quadruple $\{a, x, y, z\}$ in B with $a \neq v$, and so since $3 + 3 + 3 + 3 \equiv 0 \pmod 3$, $\{(a, 3), (x, 3), (y, 3), (z, 3)\}$ is a Type 2 quadruple, so is also in B'. □

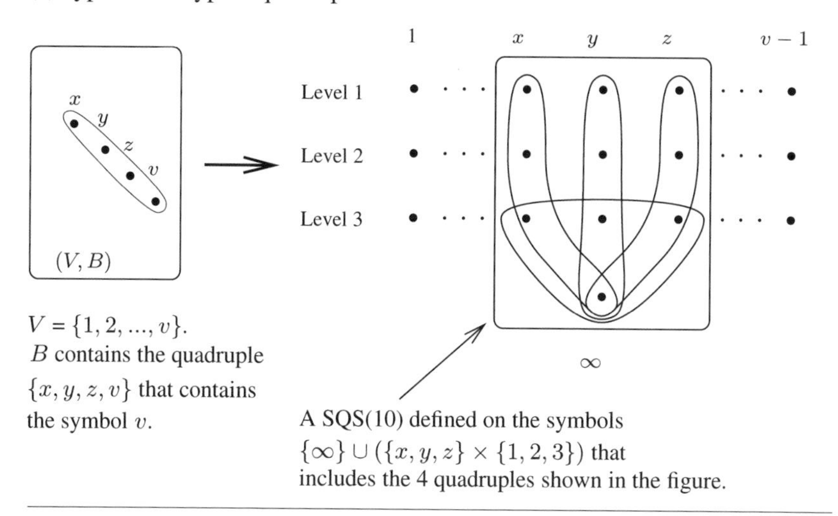

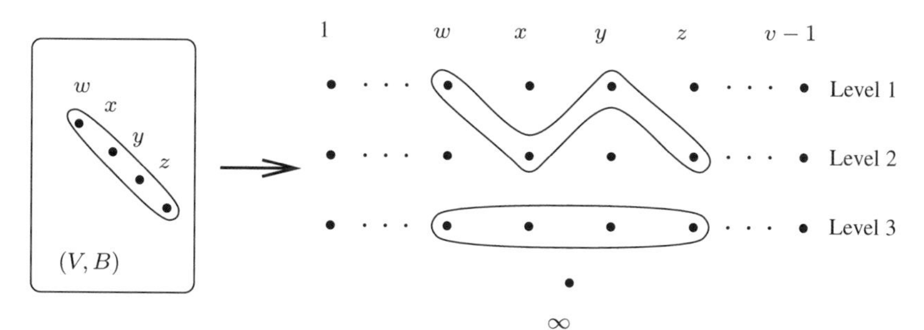

Figure 7.3: *The $(3v - 2)$ Construction.*

Exercises

7.2.7 We can use The $(3v-2)$ Construction to define a $SQS(22)$, so $v = 8$. We need a $SQS(v = 8)$, and a $SQS(10)$ for the Type 1 quadruples. We can use the $SQS(8)$ $(\{1, 2, \ldots, 8\}, B)$ defined in Example 7.1.3 by arbitrarily renaming the symbols, say renaming $000, 001, 010, 011,$ $100, 101, 110$ and 111 with $1, 2, \ldots, 8$ respectively. We can use the

$SQS(10)$ in Example 7.1.4, but in this case we have to be careful to rename the symbols in the $SQS(10)$: The $(3v-2)$ Construction requires that the $SQS(10)$ used for the Type 1 quadruples contains the four quadruples $\{\infty, (x,1), (x,2), (x,3)\}$, $\{\infty, (y,1), (y,2), (y,3)\}$, $\{\infty, (z,1), (z,2), (z,3)\}$ and $\{\infty, (x,3), (y,3), (z,3)\}$. (By Exercise 7.1.11 this is possible.) For example, we can rename the symbols $\{1,2\}$, $\{2,3\}, \{1,3\}, \{2,4\}, \{3,4\}, \{1,4\}, \{1,5\}, \{2,5\}, \{3,5\}, \{4,5\}$ in Exercise 7.1.4 with $(x,1), (y,1), (z,1), (x,2), (y,2), (z,2), (x,3), (y,3)$, $(z,3)$ and ∞ respectively, for each block $\{x, y, z, v = 8\}$ in B with $x < y < z$.

Using the above renaming scheme, The $(3v-2)$ Construction produces a $SQS(22)$ (V', B').

(a) Find the quadruple in B' containing the following 3-element sets:

(i) $\{\infty, (2,1), (4,3)\}$
(ii) $\{(2,1), (2,3), (7,2)\}$
(iii) $\{(3,1), (5,1), (6,3)\}$
(iv) $\{(1,1), (2,1), (7,1)\}$
(v) $\{(2,1), (3,3), (5,3)\}$

(b) Observe that this $SQS(22)$ (V', B') contains an $SQS(8)$ defined on the symbols $\{\infty\} \cup (\{1, 2, \ldots, 7\} \times \{3\})$.

7.2.8 Show that in The $(3v-2)$ Construction, the number of

Type 1 quadruples is $27(v-1)(v-2)/6$,

Type 2 quadruples is $27(v(v-1)(v-2)/24 - v(v-1)/6)$, and

Type 3 quadruples is $v-1$.

Deduce that $(3v-2)(3v-3)(3v-4)/24$ quadruples occur in B'. (Hint: To count the number of Type 1 quadruples, use Exercise 7.1.8.)

7.3 The Stern and Lenz Lemma

The lemma of Stern and Lenz [22] enabled Lenz [15] to drastically simplify a construction of SQS's of Alan Hartman [9, 10]. The Lenz version of this construction is presented in Section 7.4. However, we devote this section solely to the Stern and Lenz Lemma, because it is so useful in the construction of many designs when using difference methods.

We begin with some graph theoretical definitions. A *component* in a graph G is a maximal connected subgraph of G. A graph is k-*regular* if all vertices have degree k. A k-*cycle* is defined to be the graph with vertex set $\{v_1, \ldots, v_k\}$ and edge set $\{\{v_i, v_{i+1}\} | 1 \leq i \leq k-1\} \cup \{\{v_1, v_k\}\}$, and denote this k-cycle by $(v_1, v_2, \ldots, v_k)$. If H is a graph with vertex set V and edge set E, and if $A \subseteq V$, then the subgraph of H *induced* by A is the graph with vertex set A and edge set $\{\{u, v\} | u \in A, v \in A, \{u, v\} \in E\}$. We denote this induced subgraph by $H[A]$.

Let G be a graph with g vertices. The *difference* of the edge $e = \{u, v\}$ in G, named so that $u < v$, is defined to be $v - u$ or $g - (v - u)$, whichever is smaller; we denote the difference of $e = \{u, v\}$ by $D_g(u, v)$, or by $D(u, v)$ if the value of g is clear. So if $e = \{u, v\}$ with $u < v$ then

$$D(e) = D(u, v) = \min\{v - u, g - (v - u)\}$$

(where it is important to remember that we are assuming that $u < v$).

Example 7.3.1 Let G be a graph defined on the 9 vertices in $\{0, 1, \ldots, 8\}$, so $g = 9$. Then for example, if e is the edge $\{0, 7\}$ then $u = 0$ and $v = 7$ (so $u < v$) and therefore $D(0, 7) = \min\{7 - 0, 9 - (7 - 0)\} = 2$. Similarly $D(3, 6) = \min\{6 - 3, 9 - (6 - 3)\} = 3$, and $D(5, 8) = \min\{8 - 5, 9 - (8 - 5)\} = 3$.

With this definition, it is clear that $D(u, v) \leq \lfloor g/2 \rfloor$. Notice also that the number of edges of difference d in the complete graph K_g is g if $d < g/2$, and is $g/2$ if $d = g/2$ (see Exercise 7.3.8).

The following graph will be very important to us. For any subset $D \subseteq \{1, 2, \ldots, \lfloor g/2 \rfloor\}$, define $G(D, g)$ to be the graph with vertex set $\{0, 1, \ldots, g - 1\}$ and edge set consisting of all edges having a difference in D; that is, the edge set of $G(D, g)$ is $\{\{u, v\} | \, D(u, v) \in D\}$.

Example 7.3.2

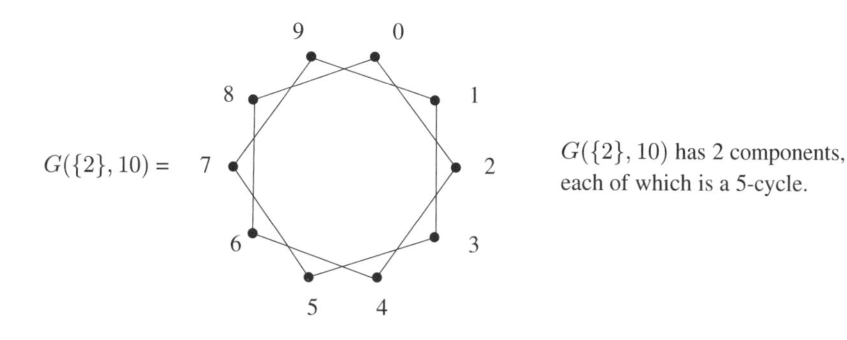

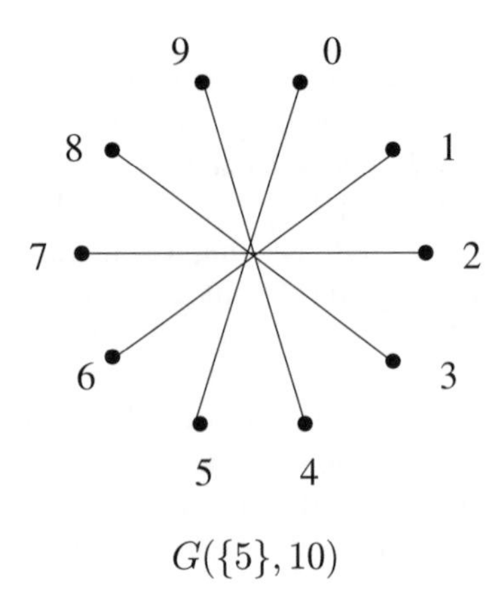

$G(\{5\}, 10)$

$G(\{5\}, 10)$ has 5 components,
each of which is a K_2.

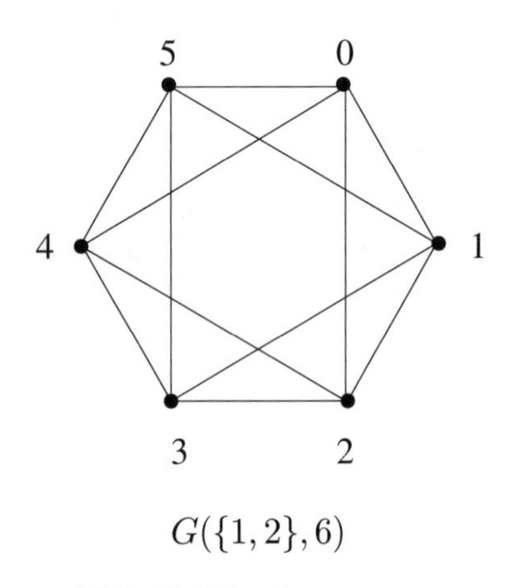

$G(\{1, 2\}, 6)$

$G(\{1, 2\}, 6)$ has 1 component.

The Stern and Lenz Lemma is concerned with finding a 1-factorization of $G(D, g)$. Before stating and proving this result, it is worth noting the following.

Lemma 7.3.3 *If h is the greatest common divisor of $D \cup \{g\}$ (or more briefly, $gcd(D \cup \{g\})$), then $G(D, g)$ consists of h components, each of which is isomorphic to $G(\{d/h | d \in D\}, g/h)$.*

Proof Since h divides g, we can write $g = g_1 h$. Then we can partition the vertex set $V = \{0, 1, \ldots, g-1\}$ of $G(D, g)$ into h sets $S_0, S_1, \ldots, S_{h-1}$, where

$$S_i = \{i, i+h, i+2h, \ldots, i+(g_1-1)h\}.$$

Let G_i be the subgraph of G induced by the vertex set S_i. We will now see that $G_0, \ldots, G_{h-1}$ are the h components of $G(D, g)$.

Suppose that $v = i + xh$ is a vertex in S_i, and suppose that e is any edge incident with v in $G(D, g)$. Then e is an edge of difference d, for some $d \in D$. Therefore h divides d, so we can write $d = d_1 h$. Since e has difference d, e joins v to either the vertex $v + d = (i + xh) + d_1 h = i + (x + d_1)h$ which is a vertex in S_i, or the vertex $v - d = i(x - d_1)h$ which is also a vertex in S_i. Therefore there is no edge in $G(D, g)$ that joins a vertex in G_i to a vertex in G_j for some $i \neq j$. Also, the fact that there is no bigger common divisor of $D \cup \{g\}$ shows that for $0 \leq i \leq h - 1$, G_i is connected (see Exercise 7.3.9). So $G_0, \ldots, G_{h-1}$ are the components of $G(D, g)$.

Finally, we need to prove that G_i is isomorphic to $G(\{d/h \mid d \in D\}, g/h)$. First notice that since $g/h = g_1$, $G(\{d/h \mid d \in D\}, g/h = g_1)$ has vertex set $\{0, 1, \ldots, g_1 - 1\}$. Also, for each $d \in D$ we can write $d = d_1 h$, so $\{d/h \mid d \in D\} = \{d_1 \mid d \in D\}$. So now define the function $f : S_i \to \{0, 1, \ldots, g - 1\}$ by $f(i + xh) = x$. Since we have seen that any edge in G_i is of the form $e = \{i + xh, i + (x \pm d_1)h\}$, f maps this edge e to $\{x, x \pm d_1\}$. Therefore it easily follows that f is an isomorphism from G_i to $G(\{d/h \mid d \in D\}, g/h)$ as required. □

Corollary 7.3.4 *$G(\{d\}, g)$ consists of $h = gcd(\{d, g\})$ components, and each component is*

(1) a (g/h)-cycle if $d \neq g/2$, and

(2) K_2 if $d = g/2$.

Example 7.3.5 (a) The graph $G(\{2\}, 10)$ in Example 7.3.2 consists of $h = gcd(\{2, 10\}) = 2$ components, each of which is a (g/h)-cycle $=$ 5-cycle.

(b) The graph $G(\{5\}, 10)$ in Example 7.3.2 consists of $h = gcd(\{5, 10\}) = 5$ components, each of which is K_2.

(c) The graph $G(\{1, 2\}, 6)$ in Example 7.3.2 consists of $gcd(\{1, 2, 6\}) = 1$ component.

(d) The graph $G(\{3\}, 12)$ consists of $h = gcd(\{3, 12\}) = 3$ cycles of length $g/h = 4$, namely $(0, 3, 6, 9)$, $(1, 4, 7, 10)$ and $(2, 5, 8, 11)$. Notice that each component is a 4-cycle, as is the graph $G(\{3/h\}, 12/h) = G(\{1\}, 4)$.

(e) The graph $G(\{6, 8, 10\}, 24)$ consists of $h = gcd(\{6, 8, 10, 24\}) = 2$ components, each of which is isomorphic to the graph $G(\{d/h \mid d \in D\}, g/h) = G(\{3, 4, 5\}, 12)$ (see Exercise 7.3.7). This graph is regular of degree 6.

(f) The graph $G(\{3, 5\}, 10)$ consists of $gcd(\{3, 5, 10\}) = 1$ component and contains 15 edges.

Exercises

7.3.6 Find $G(\{4, 5, 6\}, 12)$. Does this graph have a 1-factorization?

7.3.7 Check Example 7.3.5(e) by drawing all the edges joining two even vertices in $G(\{6,8,10\},24)$. Then draw the edges which join two odd vertices in $G(\{6,8,10\},24)$. Finally, what is $G(\{3,4,5\},12)$?

7.3.8 Show that the number of edges in $G(\{d\},g)$ is g if $d < g/2$, and is $g/2$ if $d = g/2$.

7.3.9 Show that G_i in the proof of Lemma 7.3.3 is connected. (Hint: Show that if G_i is disconnected then you can find $h' > h$ such that h' divides each element of $D \cup \{g\}$.)

The following proof uses some graph theory, which we introduce now. A *proper k-edge-coloring* of a graph $G = (V, E)$ is a function $f : E \to C$ where C is a set of size k (usually $C = \{1, 2, \ldots, k\}$) such that if two edges $e_1 \neq e_2$ are adjacent then $f(e_1) \neq f(e_2)$. You can think of $f(e)$ as being the "color" that f assigns to the edge e; then a proper k-edge-coloring makes sure that each edge receives one of the "colors" $1, 2, \ldots, k$, and that if two edges are adjacent then they receive *different* colors.

Example 7.3.10

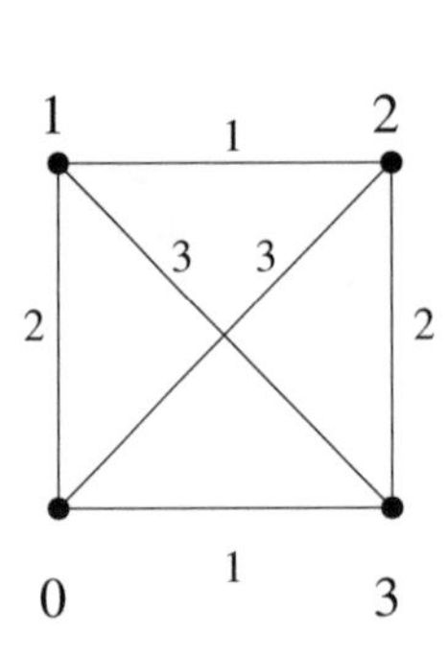

A proper 3-edge-coloring of K_4.

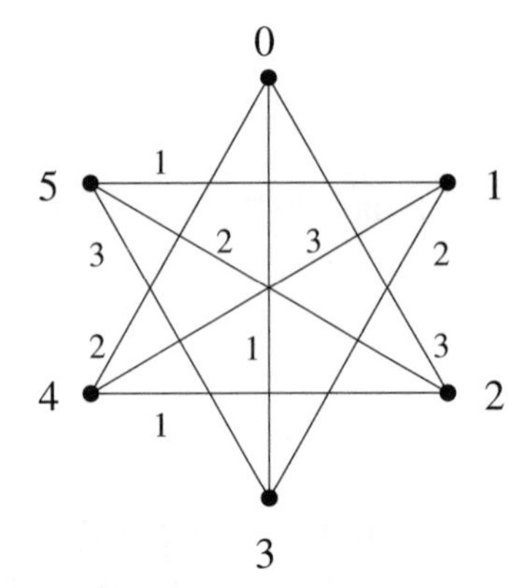

A proper 3-edge-coloring of $G(\{2,3\},6)$.

So what do edge-colorings have to do with 1-factorizations? A little reflection shows that a graph G has a 1-factorization if and only if G is k-regular and has a proper k-edge-coloring. In each graph of Example 7.3.10, the edges colored i form a 1-factor F_i, and $\{F_1, F_2, F_3\}$ is a 1-factorization. We will also use a famous result of Vizing [25].

Theorem 7.3.11 *(Vizing) A simple graph G, with k being the maximum degree among all vertices in G, has a proper $(k+1)$-edge-coloring.*

Of course, a proper $(k+1)$-edge-coloring of a k-regular graph G does not produce a 1-factorization of G, but this is a crucial ingredient in the proof of the Stern and Lenz Lemma.

Lemma 7.3.12 (Stern and Lenz) *If D contains an element d where $g/gcd(\{d, g\})$ is even, then $G(D, g)$ has a 1-factorization.*

Remark 7.3.13 If no such element $d \in D$ exists, then $G(D, g)$ does not have a 1-factorization (see Exercise 7.3.16).

Proof Suppose that $D = \{d, d_1, \ldots, d_k, \hat{d}_1, \ldots, \hat{d}_\ell\}$, where $g/gcd(\{d_i, g\})$ is odd for $1 \leq i \leq k$, $g/gcd(\{\hat{d}_i, g\})$ is even for $1 \leq i \leq \ell$, and of course $g/gcd(\{d, g\})$ is required to be even by the statement of the lemma. Then from Corollary 7.3.4, since $g/gcd(\{\hat{d}_i, g\})$ is even for $1 \leq i \leq \ell$, each component of $G(\{\hat{d}_i\}, g)$ is a cycle of even length, so the edges in each cycle can alternately be placed in $\hat{F}_{1,i}$ and $\hat{F}_{2,i}$ to form two 1-factors. Therefore, we can focus our attention on finding a 1-factorization of $G(\{d, d_1, \ldots, d_k\}, g)$.

Let $h = gcd\{d, d_1, d_2, \ldots, d_k, g\}$. From Lemma 7.3.3, each component of $G(\{d, d_1, \ldots, d_k\}, g)$ is isomorphic to $G(\{d/h, d_1/h, \ldots, d_k/h\}, g/h)$, and of course a 1-factorization of each component gives a 1-factorization of the whole graph. Therefore, if we define $d' = d/h$, $d'_i = d_i/h$ for $1 \leq i \leq k$, $g' = g/h$, and $D' = \{d', d'_1, d'_2, \ldots, d'_k\}$, then we can focus our attention even further, and simply find a 1-factorization of $G(D', g')$. Equivalently, we will find a proper Δ-edge-coloring of $G(D', g')$, where Δ is the degree of each vertex in $G(D', g')$.

Clearly, since h is the greatest common divisor of $\{d, d_1, d_2, \ldots, d_k, g\}$, we have that

$$h' = gcd(D' \cup \{g'\}) = 1.$$

Also

$$g'/gcd(\{d', g'\}) = (g/h)/gcd(\{d/h, g/h\}) = g/gcd\{d, g\}$$

which we know is even, and

$$g'/gcd\{d'_i, g'\} = (g/h)/gcd(\{d'_i/h, g/h\}) = g/gcd(d_i, g)$$

which we know is odd.

The fact that $G(D', g')$ has a 1-factorization depends heavily upon the parity of each of d', d'_i and g', so we find that now.

Clearly g' is even because $g'/gcd(\{d', g'\})$ is even. Also, since $g'/gcd(\{d'_i, g'\})$ is odd and since g' is even, $gcd(\{d'_i, g'\})$ is even so d'_i must be even for $1 \leq i \leq k$. Finally, since $gcd(D' \cup \{g'\}) = 1$ and since we have just seen that $g', d'_1, \ldots, d'_k$ are all even, d' must be odd.

Notice that by Corollary 7.3.4, if $d' = g'/2$ then $G(D', g')$ is regular of degree $\Delta = 2k+1$, and if $d' < g'/2$ then $G(D', g')$ is regular of degree $\Delta = 2k+2$ (d'_i cannot equal $g'/2$ because $g'/gcd(\{d'_i, g'\})$ is odd).

We are now ready to find a 1-factorization of $G(D', g')$. First we consider the subgraph $H = G(D' \backslash \{d'\}, g')$. Clearly H is regular of degree $2k$. Let $A = \{0, 2, 4, \ldots, g' - 2\}$ and $B = \{1, 3, 5, \ldots, g' - 1\}$. Since d_i' is even for $1 \leq i \leq k$, each edge in H is either in $H[A]$ or is in $H[B]$; that is, H contains no edges joining a vertex in A to a vertex in B. Also, $H[A]$ and $H[B]$ are both regular of degree $2k$. Furthermore, since d' is odd, $\{u, v\}$ is an edge in $H[A]$ if and only if $\{u + d', v + d'\}$ is an edge in $H[B]$.

By Vizing's Theorem, since H is $2k$-regular, the edges in $H[A]$ can be properly edge-colored using $2k+1$ colors, say $1, 2, \ldots, 2k+1$. We can use this edge-coloring of $H[A]$ to obtain a proper $(2k + 1)$-edge-coloring of $H[B]$: if $\{u, v\}$ is an edge in $H[A]$ colored c, then color the edge $\{u + d', v + d'\}$ in $H[B]$ with c. Clearly this is a proper coloring of the edges of $H[B]$.

Finally, it remains to add the edges of difference d' to H and color them so that we obtain a proper edge-coloring of $G(D', g')$. For each vertex $u \in A$, since u has degree $2k$ and since we are using $2k + 1$ colors to color the edges of H, there exists a color that does not occur on any edge in H that is incident with u; say this color is $c(u)$. By the method used to edge-color $H[B]$, the colors on the edges in $H[A]$ incident with u are precisely the same as the colors on the edges in $H[B]$ that are incident with $u + d'$; so $c(u)$ cannot occur on any edge incident with $u + d'$ in $H[B]$. Therefore, we can color the edge $\{u, u + d'\}$ in $G(D', g')$ with $c(u)$ for each $u \in A$.

If $d' = g'/2$, then this is a proper $\Delta = (2k+1)$-edge-coloring of $G(D, g')$, and hence is a 1-factorization of $G(D', g')$ as required. If $d' < g'/2$ then it still remains to color the edges $\{u, u - d'\}$ for each $u \in A$; but these can each be colored with a new color $2k + 2$. Then we have a proper $\Delta = (2k + 2)$-edge-coloring of $G(D', g')$.

Thus a 1-factorization of $G(D', g')$ has been produced, so we have a 1-factorization of $G(D, g)$. □

Example 7.3.14 The 1-factorization of $G(\{2, 3\}, 6)$ corresponding to the proper 3-edge-coloring of this graph in Example 7.3.10 was produced using the proof of Lemma 7.3.12, as the following discussion shows.

In this case $g = 6$, $d = 3$ and $d_1 = 2$ and so $gcd(\{d, d_1, g\}) = 1$ (using the notation in the proof of Lemma 7.3.12). Therefore $G(D, g)$ is connected, and so $D' = D$ and $g' = g$.

The graph $H = G(D \backslash \{d\}, g) = G(\{2\}, 6)$ consists of two 3-cycles, one defined on the vertex set $A = \{0, 2, 4\}$, and the other on the vertex set $B = \{1, 3, 5\}$. The edges $\{0, 2\}$, $\{2, 4\}$ and $\{4, 0\}$ in $H[A]$ can be properly 3-edge-colored (by Vizing's Theorem), say with 3, 1 and 2 respectively. The edges $\{0 + d, 2 + d\} = \{3, 5\}$, $\{2 + d, 4 + d\} = \{5, 1\}$ and $\{4 + d, 6 + d\} = \{1, 3\}$ in $H[B]$ are therefore correspondingly colored with 3, 1 and 2 respectively. Finally, the edges $\{u, u + d\} = \{u, u + 3\}$ for each $u \in A$ are colored with the only color that does not occur on any edge incident with u.

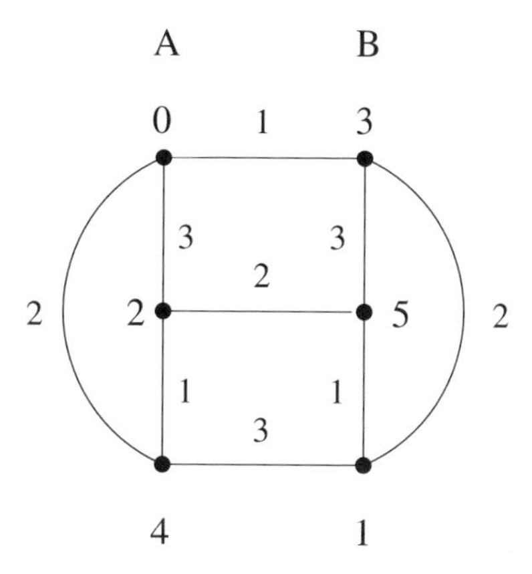

Example 7.3.15 We can apply the proof of The Stern and Lenz Lemma to find a 1-factorization of $G(\{3,5,6,12\},30)$ (so $g = 30$), because at least one of $g/gcd(\{3,g\}) = 10$, $g/gcd(\{5,g\}) = 6$, $g/gcd(\{6,30\}) = 5$, and $g/gcd(\{12, 30\}) = 5$ is even. So we can let $d = 3$, $d_1 = 6$, $d_2 = 12$ and $\hat{d_1} = 5$ (or alternatively, we could let $d = 5$ and $\hat{d_1} = 3$; see Exercise 7.3.18). The following steps are depicted graphically in Figure 7.4.

1. Since (by Corollary 7.3.4) $G(\{\hat{d_1}\},g) = G(\{5\},30)$ consists of five 6-cycles, the edges in each 6-cycle can be colored alternately in 1 and 2 to form two 1-factors.

2. It remains to place the edges in $G(\{3,6,12\},30)$ into 1-factors. But $G(\{3,6,12\},30)$ consists of $h = gcd(\{3,6,12,30\}) = 3$ components, each of which is a copy of $G(\{3/h,6/h,12/h\},30/h) = G(\{1,2,4\},10)$. So we need only find a 1-factorization of $G(1,2,4\},10)$. (In the notation of the proof of Lemma 7.3.12, we now see that $D' = \{1,2,4\}$ and $d' = d/h = 1$.)

3(a). First we consider $H = G(\{2,4\},10)$, saving the edges of difference $d' = 1$ until later. The edges in H join two vertices in $A = \{0,2,4,6,8\}$, or join two vertices in $B = \{1,3,5,7,9\}$, because all edges in H have *even* difference (in this case, difference 2 or 4). Furthermore, $e = \{u,v\}$ is an edge joining two vertices in A if and only if $e + d' = \{u+d', v+d'\} = \{u+1, v+1\} = e+1$ is an edge joining two vertices in B. Give the edges in $H[A]$ a proper 5-edge-coloring with the colors $3,4,5,6$ and 7. Then, for each edge e in $H[A]$ colored say c, let $e + d' = e+1$ in $H[B]$ also be colored c. This provides H with a proper 5-edge-coloring.

Now we deal with the edges of difference $d' = 1$. These are sorted into two types:

(b) For each vertex $u \in A$, the edge $\{u, u+1\}$ of difference d' is colored with the only one of the five colors $3,4,5,6$ and 7 that does not occur on any edge incident with u in H.

(c) For each vertex $u \in A$, the edge $\{u, u-1\}$ of difference d' is colored with color 8.

This completes a proper 8-edge-coloring of $G(\{3,5,6,12\},30)$.

1.

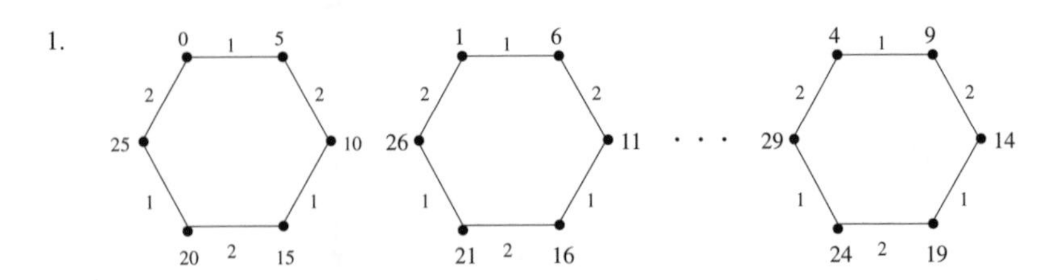

$G(\{5\}, 30)$ consists of 5 components, each being a 6-cycle, so can be properly colored.

2.

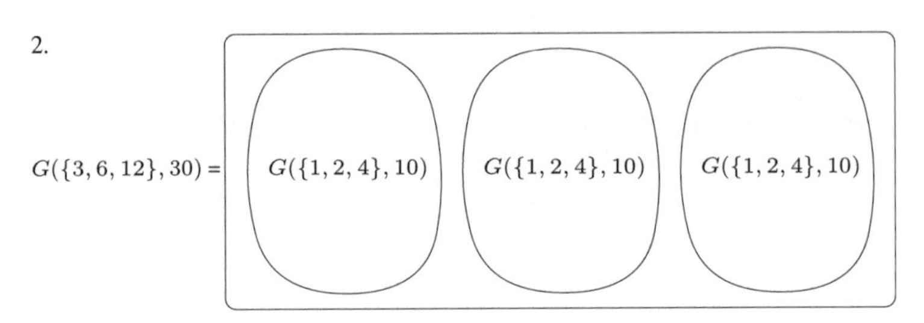

$G(\{3, 6, 12\}, 30)$ breaks up into 3 components, each of which is isomorphic to $G(\{1, 2, 4\}, 10\}$.

---

3 (a). $H = G(\{1, 2, 4\}, 10)$ is colored with the 6 colors 3,4,5,6,7 and 8 as follows.

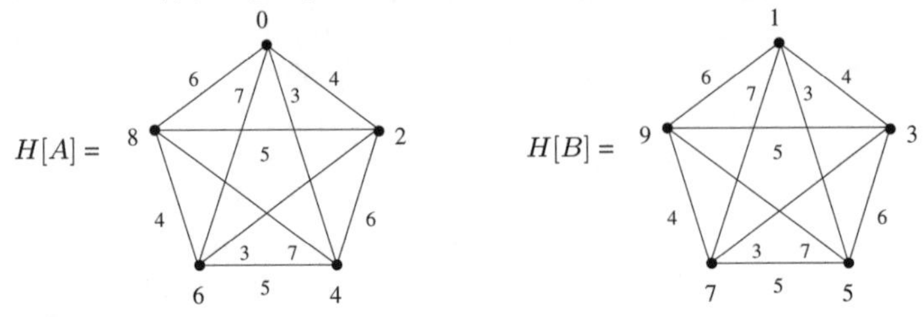

(b) $G(\{2, 4\}, 10)$ consists of the two 4-regular components above, so can be properly 5-edge-colored with the colors 3,4,5,6 and 7. Each vertex is missing exactly one of these 5 colors, and this missing color is used to color an edge of difference one in the graph below.

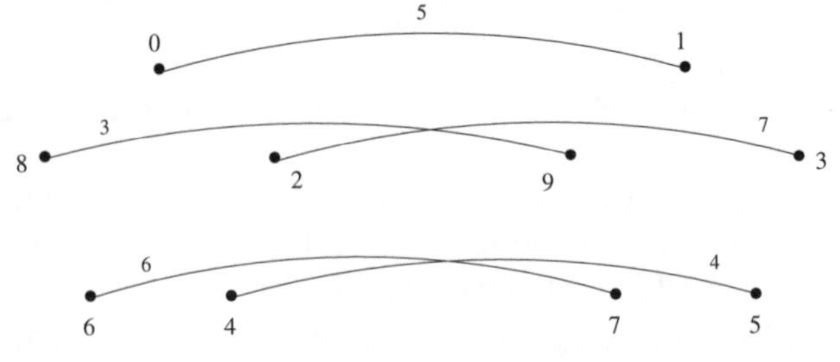

(c) $G(\{1\}, 10)$ consists of the above edges, together with 5 more edges $\{0, 9\}$, $\{2, 1\}$, $\{4, 3\}$, $\{6, 5\}$ and $\{8, 7\}$ that are each colored with 8.

Figure 7.4: *Applying the Stern and Lenz Lemma to* $G(\{3, 5, 6, 12\}, 30)$.

Exercises

7.3.16 Show that if $g/\gcd\{d, g\}$ is odd for all $d \in D$, then each component of $G(D, g)$ has an odd number of vertices. Deduce that therefore $G(D, g)$ does not have a 1-factorization.

7.3.17 Use the proof of the Stern and Lenz Lemma to find 1-factorizations of the following graphs, if a 1-factorization exists (see Exercise 7.3.16).

(a) $G(\{2, 3, 4\}, 10)$
(b) $G(\{3, 4, 5\}, 12)$
(c) $G(\{4, 6\}, 12)$
(d) $G(\{2, 4\}, 14)$
(e) $G(\{3, 6, 10\}, 30)$

7.3.18 By following Example 7.3.15, find another 1-factorization of $G(\{3, 5, 6, 12\}, 30)$ by letting $d = 5$ and $\hat{d}_1 = 3$.

7.4 The $(3v - 2u)$-Construction

In this section we present a recursive construction that starts with a $SQS(v)$ $(V = \{1, 2, \ldots, v\}, B)$ that contains a $SQS(u)$ $(V' = \{v - u + 1, \ldots, v\}, B')$ and produces a $SQS(3v - 2u)$ (V'', B''), where $V'' = \{\infty_1, \ldots, \infty_u\} \cup (\{1, 2, \ldots, v - u\} \times \{1, 2, 3\})$. This result was originally obtained by Hartman, but here we present a simplified proof due to Lenz, who made use of the Stern-Lenz Lemma.

It will simplify notation to define $g = v - u$ for the rest of this chapter. Notice that if $1 \leq a \leq g$ then a naturally corresponds to 3 symbols in V'', namely $(a, 1), (a, 2)$ and $(a, 3)$; and if $g + 1 \leq a \leq v$ then symbol a in the $SQS(v)$ naturally corresponds to one symbol in $V,''$ namely symbol ∞_{a-g}. This correspondence is formalized by defining

$$\phi_i(a) = \begin{cases} (a, i) & \text{if } 1 \leq a \leq g, \\ \infty_{a-g} & \text{if } g + 1 \leq a \leq v, \end{cases}$$

for $1 \leq i \leq 3$, and is depicted in Figure 7.5. So for example, $\phi_2(1) = (1, 2)$, and $\phi_2(v - 1) = \infty_{u-1}$. Notice that if $a > g$ then $\phi_1(a) = \phi_2(a) = \phi_3(a) = \infty_{a-g}$, reflecting the fact that in this case $a \in V$ corresponds to just one symbol $\infty_{a-g} \in V''$.

In the following discussion, as usual we refer to (a, i) as occurring on *level* i. Also, if we say a symbol occurs on level i then clearly we mean that the symbol is of the form (a, i) and is *not* one of $\infty_1, \ldots, \infty_u$. Symbols $\infty_1, \ldots, \infty_u$ are referred to as *infinite* vertices.

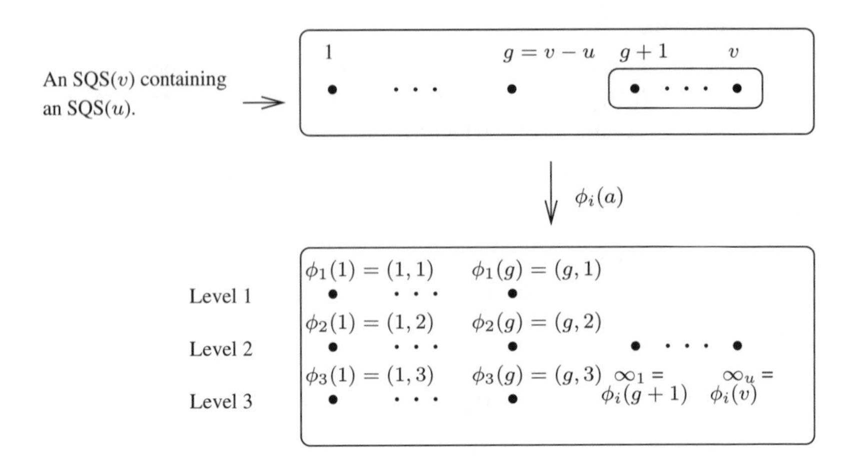

Figure 7.5: *Symbols in the $SQS(3v - 2u)$ corresponding to symbols in the $SQS(v)$.*

To simplify the notation and to make the construction easier to understand, we begin by partitioning the set of all triples on the set $V'' = (\{1, 2, \ldots, v-u\} \times \{1, 2, 3\}) \bigcup \{\infty_1, \ldots, \infty_u\}$ into 4 sets: T_1, T_2, T_3 and T_4. We will then define 4 types of quadruples Q_1, Q_2, Q_3 and Q_4 which are then cleverly combined in The $(3v - 2u)$-Construction so that these quadruples cover each of the triples exactly once. Notice that since v and u are even, so is g. We can define four types of triples: T_1, T_2, T_3 and T_4 as follows.

Type 1. Let

$$T_1 = \{\{\phi_i(x), \phi_i(y), \phi_i(z)\} | \{x, y, z\} \subseteq \{1, 2, \ldots, v\}, x \neq y \neq z \neq g, 1 \leq i \leq 3\}.$$

So the triples in T_1 are precisely all the triples that do not contain 2 symbols that are on different levels.

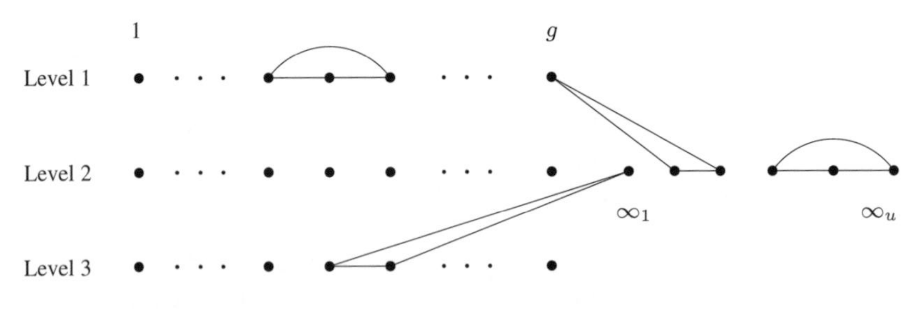

Type 2. T_2 is itself partitioned into u sets $T_2(\ell)$ for $1 \leq \ell \leq u$, where

$$T_2(\ell) = \{\{\infty_\ell, (x, i), (y, j)\} | 1 \leq x, y \leq g, 1 \leq i < j \leq 3\}.$$

So the triples in $T_2(\ell)$ are the triples that contain ∞_ℓ and contain any two symbols that are on different levels. Two such triples are shown below.

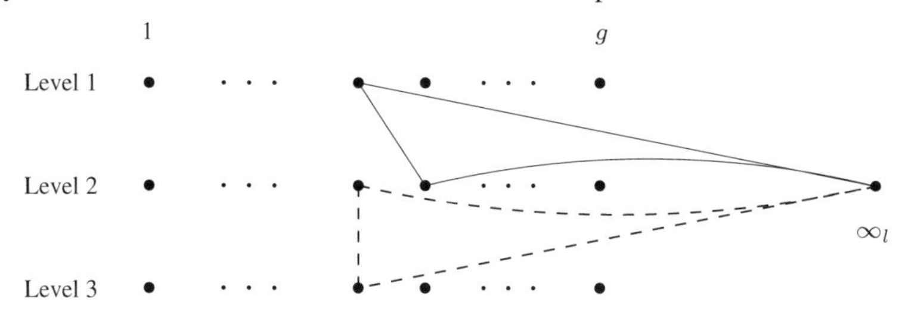

Type 3. T_3 is partitioned into g sets $T_3(a)$ for $0 \leq a \leq g-1$, where

$$T_3(a) = \{\{(x,1),(y,2),(z,3)\} | x+y+z \equiv a \text{ (mod g)}\}.$$

So the triples in $T_3(a)$ are the triples that contain any three symbols that are on three different levels whose first coordinates add to a modulo g. Two such triples are shown below.

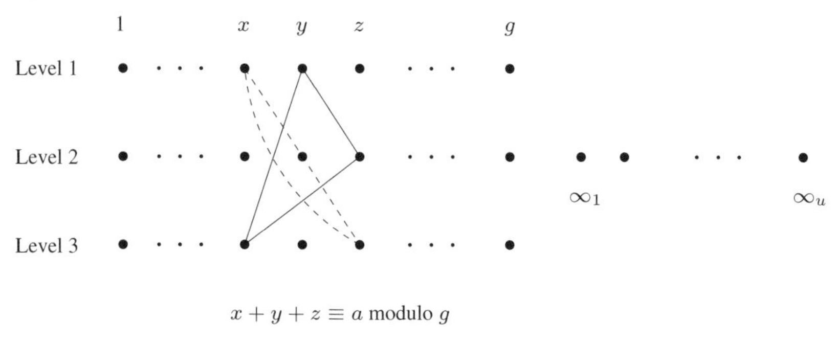

Type 4. T_4 is partitioned into $g/2$ sets $T_4(d)$ for $1 \leq d \leq g/2$, where

$$T_4(d) = \{\{(x,i),(x+d,i),(y,j)\} | 1 \leq x,y \leq g, 1 \leq i,j \leq 3, i \neq j\}.$$

So triples in $T_4(d)$ contain two symbols that are on the same level and are joined by an edge of difference d modulo g, and any third symbol that is on a different level. Two such triples are shown below.

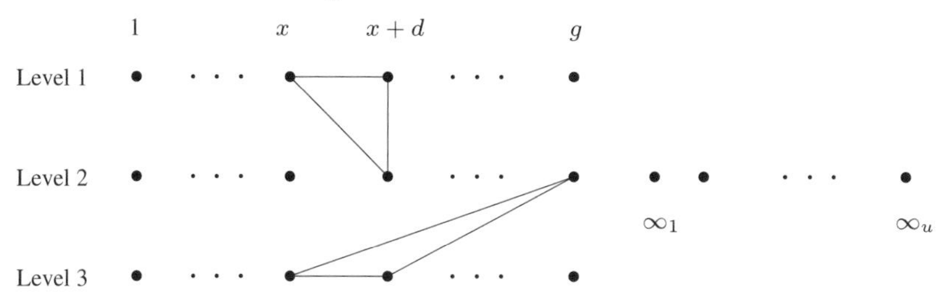

Lemma 7.4.1 *The set of triples of* $(\{1,2,\ldots,g\}\times\{1,2,3\})\bigcup\{\infty_1,\ldots,\infty_u\}$ *is partitioned by*

$$T = T_1 \cup (\bigcup_{1\le \ell\le u} T_2(\ell)) \cup (\bigcup_{1\le a\le g} T_3(a)) \cup (\bigcup_{1\le d\le g/2} T_4(d)).$$

Proof This is easy to show (see Exercise 7.4.5). □

When partitioning these triples into quadruples, it will help to know how many triples are in each set.

Lemma 7.4.2 *The number of triples*

1. *in* T_1 *is* $3\binom{g}{3} + 3\binom{g}{2}u + 3g\binom{u}{2} + \binom{u}{3}$,
2. *in* $T_2(\ell)$ *is* $3g^2$,
3. *in* $T_3(a)$ *is* g^2,
4. *in* $T_4(d)$ *is* $6g^2$ *if* $d < g/2$ *and is* $3g^2$ *if* $d = g/2$, *and*
5. *in* T *of Lemma 7.4.1 is* $\binom{3v-2u}{2}$.

Proof See Exercise 7.4.4. □

Exercises

7.4.3 For each of the following triples t, decide which of $T_1, T_2(\ell), T_3(a)$ or $T_4(d)$ contains t, and find the value of ℓ, a or d if t is in $T_2(\ell)$, $T_3(a)$ or $T_4(d)$ respectively. In each case, assume that $g = 14$.

(a) $t = \{(1,1),(4,2),(5,3)\}$
(b) $t = \{(1,1),(5,3),\infty_3\}$
(c) $t = \{(3,1),(5,1),\infty_4\}$
(d) $t = \{(2,2),(6,2),(7,1)\}$
(e) $t = \{(1,2),(9,2),(11,3)\}$
(f) $t = \{(4,1),(5,1),(12,1)\}$
(g) $t = \{\infty_1,\infty_3,\infty_4\}$
(h) $t = \{(8,3),\infty_1,\infty_2\}$
(i) $t = \{(3,1),(5,2),(11,3)\}$
(j) $t = \{(8,2),(10,2),(10,3)\}$

7.4.4 Prove Lemma 7.4.2.

7.4.5 Show that the triples in $(\{1,2,\ldots,g\}\times\{1,2,3\})\cup\{\infty_1,\ldots,\infty_u\}$ are partitioned by the sets $T_1, T_2(\ell), T_3(a)$ and $T_4(d)$ (see Lemma 7.4.1); you may want to use Lemma 7.4.2 and Exercise 7.4.4.

We now proceed to define four types of quadruples: Q_1, Q_2, Q_3 and Q_4. Each of these sets of quadruples in $(\{1,2,\ldots,g\}\times\{1,2,3\})\cup\{\infty_1,\ldots,\infty_u\}$ is carefully defined to contain each of the triples in some of the sets $T_1, T_2(\ell), T_3(a)$ and $T_4(d)$ defined above. (We will show which sets of triples each set of quadruples contains in Lemma 7.4.7.) Since these quadruples are being defined for us for use in The $(3v-2u)$-Construction, we will assume that $(\{1,\ldots,v\},B)$ is a $SQS(v)$ containing an $SQS(u)$ on the symbols $\{g+1 = v-u+1,\ldots,v\}$. These quadruples in B are used in defining Q_1.

Type 1. Let

$$Q_1 = \{\{\phi_i(x), \phi_i(y), \phi_i(z), \phi_i(w)\} | \{x, y, z, w\} \in B, 1 \leq i \leq 3\}.$$

So each quadruple in the $SQS(u)$ corresponds to one quadruple of Type 1 defined on 4 infinite vertices, and each other quadruple in the $SQS(v)$ corresponds to three quadruples of Type 1.

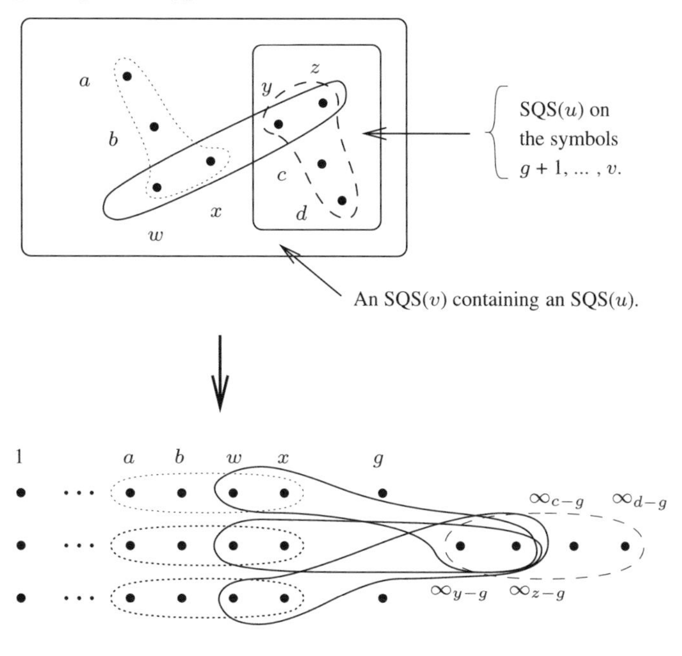

Type 2. For $0 \leq a \leq g-1$ and for $1 \leq \ell \leq u$, let

$$Q_2(a, \ell) = \{\{\infty_\ell, (x, 1), (y, 2), (z, 3)\} | x + y + z \equiv a \text{ (mod g) }\}.$$

So each quadruple in $Q_2(a, \ell)$ contains ∞_ℓ, and 3 more symbols that occur on different levels whose first coordinates add to a modulo g. Two quadruples in $Q_2(a, \ell)$ are shown below.

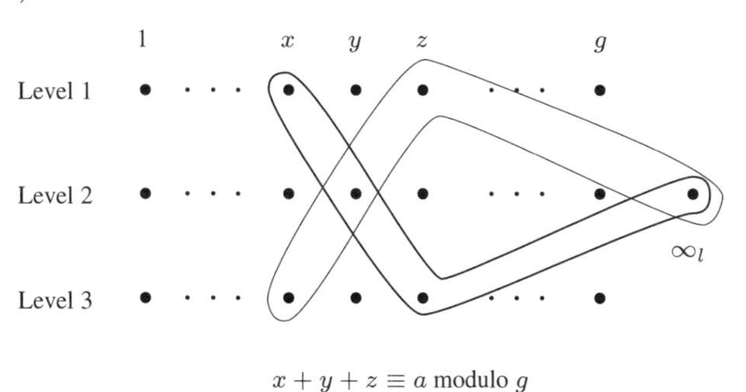

$x + y + z \equiv a$ modulo g

Type 3. For $1 \leq d < g, d \neq g/2$ and for each $\{a_1, a_2, a_3\} \subseteq \{0, 1, \ldots g-1\}$ (a_1, a_2 and a_3 need not be all different), define

$$Q_3(a_1, a_2, a_3; d) = \{\{(x, i), (x+d, i), (y, i+1), (z, i+2)\} | 1 \leq i \leq 3,$$

$$x + y + z \equiv a_i \text{ (mod g) }\},$$

reducing the first and second components modulo g and 3 respectively. (Not all of x, y and z need be different.) Two quadruples in $Q_2(a_1, a_2, a_3; d)$ are shown below.

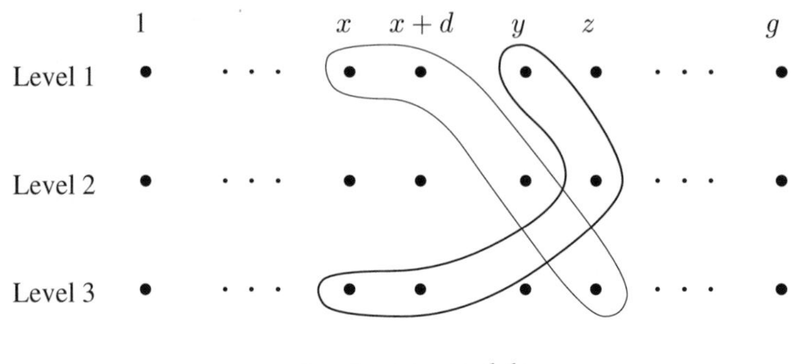

$x + y + z \equiv a$ modulo g

Type 4. For each $D \subseteq \{1, 2, \ldots, g/2\}$ that contains $g/2$, let $\{F_1, \ldots, F_t\}$ be a 1-factorization of $G(D, g)$ (see Section 7.3), and let

$$Q_4(D) = \{\{(x, i), (y, i), (z, j), (w, j)\} | 1 \leq i < j \leq 3, \{x, y\} \in F_k,$$

$$\{z, w\} \in F_k, 1 \leq k \leq t\}.$$

So each quadruple in $Q_4(D)$ contains two symbols on one level and two symbols on another level. Two quadruples in $Q_4(D)$ are shown below.

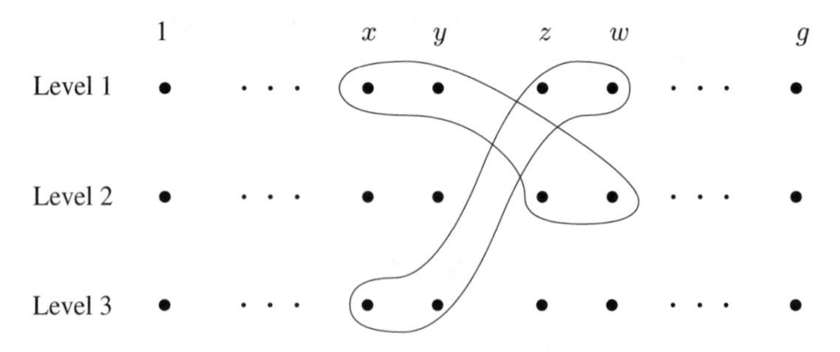

We will now look closely at the triples contained in these sets of quadruples. We begin by counting the number of triples in the quadruples in each set.

Lemma 7.4.6 *The number of triples occurring in quadruples*

1. *in* Q_1 *is* $v(v-1)(v-2)/2 - u(u-1)(u-2)/3$,
2. *in* $Q_2(a, \ell)$ *is* g^2,
3. *in* $Q_3(a_1, a_2, a_3; d)$ *is* $3g^2$, *and*
4. *in* $Q_4(D)$ *is* $3g^2t/4$,

where t *is the number of* 1*-factors in the* 1*-factorization of* $G(D, g)$.

Proof (1) The number of quadruples in the $SQS(v)$ that are not in the $SQS(u)$ is $\binom{v}{3}/4 - \binom{u}{3}/4$. Each such quadruple is used to define 3 quadruples in Q_1. Each of the $\binom{u}{3}/4$ quadruples in the $SQS(u)$ is used to define 1 quadruple in Q_1. Since each quadruple contains 4 triples, the quadruples in Q_1 contain

$$4(3(\binom{v}{3}/4 - \binom{u}{3}/4) + \binom{u}{3}/4) \text{ triples.}$$

The proof for cases (2), (3) and (4) is left to Exercise 7.4.11. □

We can now find the sets of triples contained in quadruples in Q_1, Q_2, Q_3 and Q_4.

Lemma 7.4.7 *The sets of triples occurring in quadruples in* Q_1, Q_2, Q_3 *and* Q_4 *are precisely the following sets.*

1. *The triples in quadruples in* Q_1 *are the triples in* T_1.
2. *The triples in quadruples in* $Q_2(a, \ell)$ *are the triples in* $T_2(\ell)$ *and* $T_3(a)$.
3. *The triples in quadruples in* $Q_3(a_1, a_2, a_3; d)$ *are the triples in* $T_3(a_1)$, $T_3(a_2)$, $T_3(a_3)$, $T_3(a_1 + d)$, $T_3(a_2 + d)$, $T_3(a_3 + d)$ *and* $T_4(d)$.
4. *The triples in quadruples in* $Q_4(D)$ *are the triples in* $T_4(d)$, *for each* $d \in D$.

Proof From Lemmas 7.4.2 and 7.4.6, it is easy to check that in each of the cases 1 to 4, the number of triples in the quadruples in each of the sets $Q_1, \ldots, Q_4$ equals the number of triples in the corresponding set of triples. Therefore it remains, in each of the four cases, to find for each triple t in the given set of triples at least one quadruple in the corresponding set of quadruples that contains t. We proceed case by case.

1. The fact that the quadruples in Q_1 contain precisely the triples in T_1 follows immediately from B being the set of quadruples of a $SQS(v)$.
2. Consider a triple t in $T_2(\ell)$, say $t = \{\infty_\ell, (x, i), (y, j)\}, i \neq j$. Define z such that $x + y + z \equiv a$ (mod g) and define $k \in \{1, 2, 3\}$ such that $k \notin \{i, j\}$. Then $Q_2(a, \ell)$ contains the quadruple $\{\infty_\ell, (x, i), (y, j), (z, k)\}$ and this quadruple contains t.

Now consider a triple t in $T_3(a)$, say $t = \{(x,1),(y,2),(z,3)\}$. Clearly $Q_2(a,\ell)$ contains the quadruple $\{(x,1),(y,2),\ (z,3),\infty_\ell\}$, and this quadruple contains t.

3. Let $i \in \{1,2,3\}$. Consider a triple t in $T_3(a_i)$, say $t = \{(x,1),(y,2),(z,3)\}$ where $x+y+z \equiv a_i \pmod g$. If $i = 1$ then $t \subseteq \{(x,1),(x+d,1),(y,2),(z,3)\} \in Q_3(a_1,a_2,a_3;d)$. If $i = 2$ then $t \subseteq \{(x,1),(y,2),(y+d,2),(z,3)\} \in Q_3(a_1,a_2,a_3;d)$. If $i = 3$ then $t \subseteq \{(x,1),(y,2),(z,3),(z+d,3)\} \in Q_3(a_1,a_2,a_3;d)$. Therefore, the triples in $T_3(a_1), T_3(a_2)$ and $T_3(a_3)$ are all contained in quadruples in $Q_3(a_1,a_2,a_3;d)$.

   Similarly, let $i \in \{1,2,3\}$ and consider a triple t in $T_3(a_i+d)$, say $t = \{(x,i),(y,i+1),(z,i+2)\}$ where $x+y+z \equiv a_i + d \pmod g$. Then $t \subseteq \{(x,i),(x-d,i),(y,i+1),(z,i+2)\}$, and this quadruple is an element of $Q_3(a_1,a_2,a_3;d)$ since $(x-d)+y+z \equiv a_i \pmod g$. Therefore, the triples in $T_3(a_1+d), T_3(a_2+d)$ and $T_3(a_3+d)$ are contained in quadruples in $Q_3(a_1,a_2,a_3;d)$.

   Now consider a triple t in $T_4(d)$, say $t = \{(x,i),(x+d,i),(y,j)\}$ with $i \neq j$. Define z such that $x+y+z \equiv a_i \pmod g$. (It is important to notice that since we are assuming that $d < g/2$, we cannot mistake $x+d$ for x when finding z.) Also define $k \in \{1,2,3\}$ such that $k \notin \{i,j\}$. Then $t \subseteq \{(x,i),(x+d,i),(y,j),(z,k)\} \in Q_3(a_1,a_2,a_3;d)$. Therefore, the triples in $T_4(d)$ are contained in quadruples in $Q_3(a_1,a_2,a_3;d)$.

4. Consider a triple t in $T_4(D)$, say $t = \{(x,i),(x+d,i),(y,j)\}$ for some $d \in D$. In the 1-factorization of $G(D,g)$, guaranteed to exist by the Stern-Lenz Lemma, the edge $\{x,x+d\}$ occurs in a 1-factor, say F_k. The vertex y is incident with an edge in F_k, say the edge $\{y,z\}$. Then $t \subseteq \{(x,i),(x+d,i),(y,j),(z,j)\}$ which is a quadruple in $Q_4(D)$. Therefore statement 4 of the Lemma follows and the proof is complete.

□

Example 7.4.8 In the following questions, let $\{1,2,\ldots,10\}$ be the $SQS(10)$ defined in Example 7.1.4 with the edges $\{1,2\},\{1,3\},\{1,4\},\{1,5\},\{2,3\},\{2,4\},\{2,5\},\{3,4\},\{3,5\},\{4,5\}$ being renamed with $1,2,\ldots,10$ respectively. We can use the $(3v-2u)$ Construction to make a $SQS(26)$ by letting $v = 10$ and $u = 2$, so $g = 8$. Find the quadruple

(a) in Q_1 containing the triple $t = \{(2,3),(6,3),\infty_1\}$;

(b) in $Q_2(4,2)$ containing the triple $t = \{(2,1),(6,3),\infty_2\}$;

(c) in $Q_3(2,1,3;3)$ containing the triple $t = \{(3,2),(8,2),(3,3)\}$;

(d) in $Q_3(2,1,3;3)$ containing the triple $t = \{(1,1),(2,2),(8,3)\}$; and

(e) in $Q_3(2,1,3;3)$ containing the triple $t = \{(1,1),(2,2),(2,3)\}$.

(a) We have $\phi_3(2) = (2,3), \phi_3(6) = (6,3)$ and $\phi_3(9) = \infty_{9-8} = \infty_1$; so we find the quadruple in B containing the triple $\{2,6,9\}$. According to the renaming chosen for this example, $\{2,6,9\}$ corresponds to the triple $\{\{1,3\},\{2,4\},\{3,5\}\}$. So returning to Example 7.1.4, we see that the quadruple containing the edges $\{1,3\},\{2,4\}$ and $\{3,5\}$ is $\{\{1,3\}, \{2,4\}, \{3,5\}, \{1,5\}\}$ (forming a $K_3 + K_2$). According to the renaming here, $\{1,5\}$ is renamed 4, and $\phi_3(4) = (4,3)$, so the quadruple in Q_1 containing t is $\{(2,3),(4,3),(6,3),\infty_1\}$.

(b) Quadruples in $Q_2(a,\ell)$ contain one infinite symbol, and three more symbols that are all on different levels. So clearly in this case the fourth symbol (z,k) in the quadruple containing t occurs on level $k = 2$. Since we are given $Q_2(a,\ell) = Q_2(4,2)$, we now simply require z to satisfy $2+6+z \equiv a = 4 \pmod 8$, so $z = 4$. Therefore, the quadruple in $Q_2(4,2)$ containing t is $\{(2,1),(4,2),(6,3),\infty_2\}$.

(c) Here we are given $a_1 = 2$, $a_2 = 1$, $a_3 = 3$ and $d = 3$. Since $(3,2)$ and $(8,2)$ occur on the same level, we must first decide which of 3 and 8 is x, and which is $x+d$; since $8+d = 8+3 \equiv 3 \pmod{g = 8}$, we know $x = 8$ and $x+d = 3$. So now we have that $(x,i) = (8,2)$, $(x+d,i) = (3,2)$ and $(y,j) = (3,3)$. Therefore, we require (z,k) to satisfy $x + y + z = 8+3+z \equiv a_i = a_2 = 1 \pmod 8$, and $k \in \{1,2,3\}$, $k \notin \{2,3\}$. So $(z,k) = (6,1)$, and the quadruple in $Q_2(2,1,3;3)$ containing t is $\{(6,1),(3,2),(8,2),(3,3)\}$.

(d) In this case the 3 symbols in t all occur on different levels. So first we should decide on which level the fourth symbol in the quadruple in $Q_3(2,1,3;3)$ that contains t will occur. The sum of the first three coordinates of the symbols in t must be one of $a_1, a_2, a_3, a_1+d, a_2+d$ and a_3+d. Since the sum of the first coordinates of the 3 symbols in t is $1+2+8 \equiv 3 = a_3 \pmod 8$, we have that the fourth symbol occurs on level $i = 3$, so $(x,i) = (8,3)$, and $(x+d,i) = (3,3)$. Therefore, the quadruple in $Q_3(2,1,3;3)$ containing t is $\{(1,1),(2,2),(3,3),(8,3)\}$.

(e) In this case we proceed similarly to (c) above. The sum of the first coordinates of the 3 symbols in t is $1+2+2 = 5 = a_1+d$. So $i = 1$, $(x+d,i) = (1,1)$, and so $(x,i) = (6,1)$. Therefore the quadruple in $Q_3(2,1,3;3)$ containing t is $\{(1,1),(6,1),(2,2),(2,3)\}$.

Exercises

7.4.9 Continuing with Example 7.4.8, so $g = 8$, find the quadruple

(a) in Q_1 containing $\{(2,3),(3,3),(6,3)\}$,
(b) in Q_1 containing $\{(2,2),(5,2),\infty_1\}$,
(c) in $Q_2(6,1)$ containing $\{(1,1),(6,3),\infty_1\}$,
(d) in $Q_3(2,4,6;1)$ containing $\{(1,1),(7,2),(6,3)\}$,

(e) in $Q_3(2,4,6;3)$ containing $\{(2,1),(7,2),(6,3)\}$,
(f) in $Q_3(2,4,6;3)$ containing $\{(1,1),(4,1),(6,3)\}$, and
(g) in $Q_3(2,4,6;1)$ containing $\{(1,2),(8,2),(6,3)\}$.

7.4.10 Again continuing Example 7.4.8,

(a) find a 1-factorization of $G(\{2,3,4\},8)$ (see Section 7.3), then
(b) use this 1-factorization to find the quadruple in $Q_4(\{2,3,4\})$ that contains the triple

(i) $\{(1,1),(3,1),(6,3)\}$, and
(ii) $\{(1,1),(6,1),(6,3)\}$.

7.4.11 Prove cases (2), (3) and (4) of Lemma 7.4.6.

Of course we do not obtain a SQS by taking all the quadruples of all the four types. Many triples occur in more than one of these quadruples. For example, if $x+y+z \equiv b \pmod g$, then the triple $\{(x,1),(y,2),(z,3)\}$ occurs in $Q_2(b,\ell)$, $Q_3(b,a_2,a_3;d)$, $Q_3(a_1,b,a_3;d)$, and several more! So it is our job to select some of these quadruples to cover each triple exactly once.

It doesn't take long to notice a difficulty! To include each triple of Type 2, we need to select exactly u of the Type 2 quadruples: $Q_2(a_1,1), Q_2(a_2,2),\ldots,$ $Q_2(a_u,u)$ (since $Q_2(a,\ell)$ is the only set of quadruples that contains ∞_ℓ). But now we have included u of the sets of Type 3 triples, namely $T_3(a_1),\ldots,T_3(a_u)$. This leaves $g-u$ further sets of Type 3 triples which must occur in quadruples of Type 3. But $Q_3(a_i,a_j,a_k;d)$ contains 6 sets of Type 3 triples! Therefore, we need 6 to divide $g-u=v-2u$. Since each of u and v is 2 or 4 (mod 6), this only happens when $u \not\equiv v \pmod 6$. Therefore, we actually need to define two $(3v-2u)$ Constructions. So here is the first one! Recall that $D_g(u,v)$ is the difference of the edge $\{u,v\}$.

The $(3v-2u)$ Construction, $v \not\equiv u$ (mod 6). (To form a $SQS(3v-2u)$ from a $SQS(v)$ that contains a $SQS(u)$.)

Suppose that $v \not\equiv u \pmod 6$ and that there exists a $SQS(v)$ containing a $SQS(u)$. Since $v \not\equiv u \pmod 6$, and since both v and u are congruent to 2 or 4 (mod 6), it follows that $v = 6n+2u$ for some $n \geq 0$. So then $g = v-u = 6n+u$. Let $(\{1,2,\ldots,v\},B)$ be a $SQS(v)$ containing a $SQS(u)$ $(\{g+1,\ldots,v\},B')$. Define a $SQS(3v-2u)$ $((\{1,2,\ldots,g\}\times\{1,2,3\})\cup\{\infty_1,\ldots,\infty_u\},B'')$ as follows.

(a) $Q_1 \subseteq B''$;

(b) $Q_2(6n+\ell,\ell) \subseteq B''$ for $1 \leq \ell \leq u$;

(c) $Q_3(3(n-i)+1, 3(n-i)+2, 3(n-i)+3; 6i-3) \subseteq B''$ for $1 \leq i \leq n$; and

(d) $Q_4(D) \subseteq B''$ for $D = \{1,2,\ldots,g/2\}\backslash\{D_g(0,6i-3) | 1 \leq i \leq n\}$.

Lemma 7.4.12 *The* $(3v - 2u)$ *Construction,* $v \not\equiv u$ *(mod 6) produces a* $SQS(3v-2u)$ *that contains a* $SQS(v)$.

Proof Using Lemma 7.4.7, each triple in T_1 is in a quadruple in (a). Each triple in T_2 is in a quadruple in (b). Each triple in $T_3(a)$ is in a quadruple in (b) for $6n+1 \le a \le g$ and is in a quadruple in (c) for $1 \le a \le 6n$ (notice that $T_3(a) = T_3(g-a)$). Each quadruple in $T_4(d)$ is in a quadruple in (c) if $d \in \{D_g(0, 6i-3) \mid 1 \le i \le n\}$ and is in a quadruple in (d) otherwise (notice that $g/2 \in D$). So by Exercise 7.4.5, a $SQS(3v-2u)$ has been produced. Clearly the quadruples in Q_1 form three $SQS(v)$s in the $SQS(3v-2u)$, each of which intersects each other in a $SQS(u)$, so the result follows. □

To show the flexibility of this construction, the following example shows how The $(3v-2u)$ Construction can be used with $v = 10$ and $u = 2$ to construct a $SQS(26)$ that not only contains a $SQS(v = 10)$ as is guaranteed to exist by Lemma 7.4.12, but also contains a $SQS(8)$.

Example 7.4.13 Let $v = 10$ and $u = 2$, so $g = 8$. Then $v \not\equiv u$ (mod 6) and there exists a $SQS(10)$ containing (trivially) a $SQS(2)$. So we can use Lemma 7.4.12 to define a $SQS(26)$ containing a $SQS(10)$. In fact, as we show below, it is possible to define our $SQS(26)$ so that it also contains a $SQS(8)$. Writing $v = 6n + 2u$ shows $n = 1$, so from The $(3v-2u)$ Construction we see the quadruples in the $SQS(26)$ are the quadruples in

(a) Q_1,

(b) $Q_2(7,1)$ and $Q_2(8,2)$,

(c) $Q_3(1,2,3;3)$, and

(d) $Q_4(\{1,2,4\})$.

To see that each triple is in one of the quadruples in one of these sets, first notice that from Lemma 7.4.1 each triple occurs in exactly one of: $T_1, T_2(1), T_2(u = 2), T_3(1), T_3(2), T_3(3), T_3(4), T_3(5), T_3(6), T_3(7), T_3(g = 8), T_4(1), T_4(2), T_4(3)$ and $T_4(g/2 = 4)$. Now, using Lemma 7.4.7 we know that: the triples in T_1 occur in quadruples in Q_1; triples in $T_2(1)$ are in quadruples in $Q_2(7,1)$; triples in $T_2(2)$ are in quadruples in $Q_2(8,2)$; triples in $T_3(1), \ldots, T_3(6)$ are in quadruples in $Q_3(1,2,3;3)$; triples in $T_3(7)$ are in quadruples in $Q_2(7,1)$; triples in $T_3(8)$ are in quadruples in $Q_2(8,2)$; triples in $T_4(1), T_4(2)$ or $T_4(4)$ are in quadruples in $Q_4(\{1,2,4\})$; and triples in $T_4(3)$ are in quadruples in $Q_3(1,2,3;3)$.

In this $SQS(26)$ it is possible to make sure that it contains a $SQS(8)$. To do this, first name the quadruples in the $SQS(10)$ so that $\{1,3,5,7\}$ is a quadruple. Then to form $Q_4(\{1,2,4\})$, obtain the 1-factorization $\{F_1, F_2, \ldots, F_7\}$ of $G(\{1,2,4\},8)$ by taking a 1-factorization of each of $G(\{1\},8), G(\{2\},8)$ and $G(\{4\},8)$ in turn (this can be done by Corollary 7.3.4 since $g/gcd\{g,d\}$ is even for each $d \in \{1,2,4\}$). Then it follows that the edges $\{1,3\}$ and $\{5,7\}$ occur in the same 1-factor, say F_1; $\{1,5\}$ and $\{3,7\}$ occur in F_2; and $\{1,7\}$ and $\{3,5\}$

occur in F_3. Therefore, the $SQS(26)$ contains a $SQS(8)$ defined on the set of symbols $\{1,3,5,7\} \times \{1,2\}$.

Example 7.4.14 Let $v = 14$ and $u = 4$. Then $v \not\equiv u \pmod 6$ and there exists a $SQS(14)$ containing (trivially) a $SQS(4)$. So we can use Lemma 7.4.12 to define a $SQS(34)$ containing a $SQS(14)$. The quadruples in the $SQS(34)$ are the quadruples in:

(a) Q_1;

(b) $Q_2(7,1), Q_2(8,2), Q_2(9,3)$ and $Q_2(0,4)$;

(c) $Q_3(1,2,3;3)$; and

(d) $Q_4(\{1,2,4,5\})$.

Exercises

7.4.15 There are many other ways of selecting quadruples of Types 1, 2, 3 and 4 to form a $SQS(3v-2u)$ than those defined in The $(3v-2)$ Construction. Find different sets of quadruples of Types 1, 2, 3 and 4 than those of Examples 7.4.13 and 7.4.14 to produce SQSs of orders 26 and 34.

7.4.16 Check explicitly that the $SQS(26)$ in Example 7.4.13 contains a $SQS(8)$ as described.

7.4.17 Following Example 7.4.13, show that the $SQS(34)$ in Example 7.4.14 can be constructed so that it contains a $SQS(8)$. (Hint: Again define a $SQS(8)$ on the set of symbols $\{1,3,5,7\} \times \{1,2\}$. To do this, obtain a 1-factorization $\{F_1, F_2, \ldots, F_7\}$ of $G(\{1,2,4,5\},10)$ in which F_1 contains $\{1,3\}$ and $\{5,7\}$, F_2 contains $\{1,5\}$ and $\{3,7\}$, and F_3 contains $\{1,7\}$ and $\{3,5\}$.)

Lemma 7.4.12 deals with the case where $v \not\equiv u \pmod 6$. In this case $g = 6n+u$ and so the triples in $T_3(a)$ for $1 \le a \le g$ can be partitioned into u sets contained in quadruples in $Q_2(a,\ell)$, and into quadruples in $Q_3(a_1,a_2,a_3;d)$, six at a time. However, if $v \equiv u \pmod 6$ then $g \equiv 0 \pmod 6$. So in such a case, after using the triples in $T_3(a)$ in quadruples in $T_2(a,\ell)$ for u values of a, the number of remaining values of a is $g-u$ which is not divisible by 6. Therefore if $v \equiv u \pmod 6$ the quadruples Q_1, Q_2, Q_3 and Q_4 are not sufficient to construct a $SQS(3v-2u)$. We overcome this problem by defining a fifth set of quadruples Q_5 to be used when $v \equiv u \pmod 6$, so $g \equiv 0 \pmod 6$. In fact Q_5 is nothing new, as it actually consists of some quadruples of Type 3 ($Q_5'(a,d)$ below), together with some quadruples of Type 4 ($Q_5''(a,d)$ below).

Type 5. Let $g \equiv 0 \pmod 6$. For $1 \le a \le g$ and for $1 \le d < g$ with $d \ne g/2$, if $a \not\equiv 0 \pmod 3$ and if $a \equiv d \pmod 3$ then define

$$\begin{aligned}
Q_5'(a,d) &= \{\{(x,i),(x+d,i),(y,i+1),(z,i+2)\}|1\leq i\leq 3,\\
&\quad y\equiv z \ (\mathrm{mod}\ 3),\ x+y+z\equiv a \ (\mathrm{mod}\ g)\},\\
Q_5''(a,d) &= \{\{(x,i),(x+d,i),(y,j),(y+d,j)\}|1\leq i<j\leq 3,\\
&\quad x\equiv y \ (\mathrm{mod}\ 3)\},\ \text{and}\\
Q_5(a,d) &= Q_5'(a,d)\bigcup Q_5''(a,d).
\end{aligned}$$

Then $Q_5(a,d)$ contains each triple in $T_3(a), T_3(a+d)$ and $T_4(d)$.

Lemma 7.4.18 *Let $g\equiv 0$ (mod 6), $a\not\equiv 0$ (mod 3), $a\equiv d$ (mod 3), $1\leq a\leq g$ and $1\leq d<g/2$. The sets of triples occurring in quadruples in $Q_5(a,d)$ are precisely $T_3(a), T_3(a+d)$ and $T_4(d)$.*

Proof $Q_5'(a,d)$ contains g^2 quadruples, since there are 3 choices for which level (x,i) is to occur on, g choices for y, and then $g/3$ choices remain for z. Similarly, $Q_5''(a,d)$ contains g^2 quadruples, since $d\neq g/2$. So the number of triples occurring in quadruples in $Q_5(a,d)$ is $4(g^2+g^2)$ which is the number of triples occurring in $T_3(a)\cup T_3(a+d)\cup T_4(d)$ (see Lemma 7.4.2). Therefore, it remains to show that each of these triples occurs in at least one of the quadruples in $Q_5(a,d)$.

Consider a triple $t=\{(x,i),(y,i+1),(z,i+2)\}\in T_3(a)$, where $1\leq i\leq 3$. Then $x+y+z\equiv a$ (mod g). Since $g\equiv 0$ (mod 3) and $a\not\equiv 0$ (mod 3), $x+y+z\not\equiv 0$ (mod 3). So exactly two of x, y and z are congruent to each other modulo 3; say $x\not\equiv y\equiv z$ (mod 3). Then t is in $\{(x,i),(x+d,i),(y,i+1),(z,i+2)\}\in Q_5'(a,d)$.

Consider a triple $t=\{(x,i),(y,i+1),(z,i+2)\}\in T_3(a+d)$. Then $x+y+z\equiv a+d$ (mod g). Since $g\equiv 0$ (mod 3), $a\equiv d$ (mod 3) and $a\not\equiv 0$ (mod 3), we see that $a+d\not\equiv 0$ (mod 3) and so $x+y+z\not\equiv 0$ (mod 3). Therefore, exactly two of x, y and z are congruent modulo 3; say $x\not\equiv y\equiv z$ (mod 3). Then t is in $\{(x-d,i),(x,i),(y,i+1),(z,i+2)\}\in Q_5'(a,d)$.

Consider a triple $t=\{(x,i),(x+d,i),(y,j)\}\in T_4(d)$. We consider three cases.

If $x+2y\equiv 0$ (mod 3), then $x\equiv y$ (mod 3). So $t\in\{(x,i),(x+d,i),(y,j),(y+d,j)\}\in Q_5''(a,d)$.

If $x+2y\equiv a$ (mod 3), then $x\not\equiv y$ (mod 3). Define z such that $x+y+z\equiv a$ (mod g). Since $x+y+z\equiv a$ (mod g), $x+2y\equiv a$ (mod 3) and $g\equiv 0$ (mod 3), we have that $y\equiv z$ (mod 3). So let $k\in\{1,2,3\}$ with $k\notin\{i,j\}$, and then $t\in\{(x,i),(x+d,i),(y,j),(z,k)\}\in Q_5'(a,d)$.

If $x+2y\equiv 2a$ (mod 3), then $x\equiv y-a\equiv y-d$ (mod 3). So $t\in\{(x,i),(x+d,i),(y-d,j),(y,j)\}\in Q_5''(a,d)$. □

Example 7.4.19 Let $g=12$. For each of the following triples t, we find the quadruple in $Q_5(5,2)$ containing t.

(a) $t = \{(1,1),(3,1),(5,2)\}$

(b) $t = \{(1,1),(3,1),(3,2)\}$

(c) $t = \{(3,1),(1,2),(3,3)\}$

Notice that we have $g \equiv 0 \pmod 6$, $5 = a \not\equiv 0 \pmod 3$, and $5 = a \equiv d = 2 \pmod 3$ as required in the definition of Type 5 quadruples.

(a) Since $(1,1)$ and $(3,1)$ occur on the same level, and since $1 + d = 3$, we have that $x = 1$ and $x + d = 3$. Then clearly $y = 5$. Following the proof of Lemma 7.4.18, we note that $x + 2y = 1 + 10 \equiv a \pmod 3$. So we define z such that $x + y + z \equiv a \pmod g$, so $z = 11$. Then choose $k \in \{1,2,3\}$ with $k \notin \{1,2\}$, so $k = 3$. So $(z,k) = (11,3)$ and the quadruple is $\{(1,1),(3,1),(5,2),(11,3)\}$.

(b) Again $x = 1$, but now $y = 3$. Following the proof of Lemma 7.4.18 we note that $x + 2y = 1 + 6 \equiv 2a \pmod 3$. So the required quadruple is $\{(1,1),(3,1),(3,2),(1,2)\}$.

(c) In this case, $x + y + z = 3 + 1 + 3 \equiv a + d \pmod g$. Since $x \equiv z$, the fourth point is $(y - d, 2) = (11, 2)$.

Exercises

7.4.20 Let $g = 12$. For each of the following triples t, find the quadruple in $Q_5(7,1)$ containing t.

(a) $t = \{(2,2),(5,3),(6,3)\}$

(b) $t = \{(2,2),(4,3),(5,3)\}$

(c) $t = \{(1,1),(4,2),(2,3)\}$

(d) $t = \{(1,1),(5,2),(2,3)\}$

(e) $t = \{(2,1),(2,2),(4,3)\}$

We can now present our second $(3v - 2u)$ Construction. The following result can easily be generalized to all $u \equiv 2 \pmod 6$ (see Exercise 7.4.24), but $u = 2$ is sufficient for our purposes.

The $(3v-2u)$ Construction, $v \equiv u = 2$ (mod 6). (To construct a $SQS(3v-2u)$ from a $SQS(v)$ that contains a $SQS(u = 2)$.)
Suppose that $v \equiv 2 \pmod 6$ and that there exists a $SQS(v)$ containing a $SQS(2)$. Since $v \equiv 2 \pmod 6$ and $u = 2$, we can write $g = 6n + 4 + u = 6n + 6$ for some $n \geq 0$. Let $(\{1,2,\ldots,v\}, B)$ be a $SQS(v)$ (trivially containing a $SQS(2)$ $(\{v-1, v\}, \phi)$). Define a $SQS(3v-4)$ $((\{1,2,\ldots,g\} \times \{1,2,3\}) \bigcup \{\infty_1, \infty_2\}, B'')$ as follows

(a) $Q_1 \subseteq B''$,

(b) $Q_2(3,1) \subseteq B'', Q_2(3n+9,2) \subseteq B''$,

(c) $Q_3(3n+3, 3n+5, 3n+7; 1) \subseteq B''$ if $n \geq 1$ and
$Q_3(3(n-i)+6, 3(n-i)+7, 3(n-i)+8; 6i-2) \subseteq B''$ for $2 \leq i \leq n$,

(d) $Q_4(D) \subseteq B''$ for $D = \{1, 2, \ldots, g/2\} \backslash (\{D_g(0, 6i-2) \mid 2 \leq i \leq n\} \cup \{1, 2, 4\})$, and

(e) $Q_5(1, 4) \subseteq B''$ and $Q_5(2, 2) \subseteq B''$.

Lemma 7.4.21 *The $(3v-2u)$ Construction with $v \equiv u = 2 \pmod 6$ produces a $SQS(3v-4)$ that contains a $SQS(v)$.*

Proof We use Lemmas 7.4.7 and 7.4.18. Each triple in T_1 is in a quadruple in (a). Each triple in T_2 is in a quadruple in (b). Each triple in $T_3(a)$ is in a quadruple in (b) for $a \in \{3, 3n+9\}$, is in a quadruple in (c) for $6 \leq a \leq 6n+6, a \neq 3n+9$, and is in a quadruple in (e) for $a \in \{1, 2, 4, 5\}$. Each triple in $T_4(d)$ is in a quadruple in (c) for each $d \in \{1, D_g(0, 6i-2) \mid 2 \leq i \leq n\}$, is in a quadruple in (e) if $d \in \{2, 4\}$, and is in a quadruple in (d) otherwise. □

Example 7.4.22 Let $v = 14$ and $u = 2$. Then $v \equiv 2 \pmod 6$ and there exists a $SQS(14)$ containing (trivially) a $SQS(2)$. So we can use The $(3v-2u)$ Construction to construct a $SQS(38)$ containing a $SQS(14)$. We have $g = v - u = 12$, and $g = 6n+6$, so $n = 1$. The quadruples in the $SQS(38)$ are the quadruples in

(a) Q_1,

(b) $Q_2(3, 1)$ and $Q_2(12, 2)$,

(c) $Q_3(6, 8, 10; 1)$,

(d) $Q_4(\{3, 5, 6\})$, and

(e) $Q_5(1, 4)$ and $Q_5(2, 2)$.

In this $SQS(38)$ it is possible to make sure that it contains a $SQS(8)$. To do so, first name the quadruples in the $SQS(14)$ so that $\{1, 4, 7, 10\}$ is a quadruple. Then obtain the 1-factorization of $G(\{3, 5, 6\}, 12)$ by taking a 1-factorization of $G(\{3, 6\}, 12)$ and of $G(\{5\}, 12)$ (see Lemma 7.3.12). It then follows that the $SQS(38)$ contains a $SQS(8)$ defined on the set of symbols $\{1, 4, 7, 10\} \times \{1, 2\}$.

Exercises

7.4.23 Find sets of quadruples Q_1, Q_2, Q_3, Q_4 and Q_5 other than those defined in Lemma 7.4.21 to produce SQSs of orders 38 and 56.

7.4.24 Modify the proof of Lemma 7.4.21 to prove that if $v \equiv u \equiv 2 \pmod 6$ (not just $u = 2$), and if there exists a $SQS(v)$ containing a $SQS(u)$, then there exists a $SQS(3v-2u)$ containing a $SQS(v)$.

7.4.25 Modify the proof of Lemma 7.4.21 to prove that if $v \equiv u \equiv 4 \pmod 6$, and if there exists a $SQS(v)$ containing a $SQS(u)$, then there exists a $SQS(3v-2u)$ containing a $SQS(v)$.

We are finally ready to settle the problem of finding the set of integers for which there exists $SQS(w)$s. From Lemma 7.1.14 we already have that either $w \equiv 2$ or 4 (mod 6) or $w = 1$.

Theorem 7.4.26 *For all $w \equiv 2$ or 4 (mod 6) or $w = 1$, there exists a $SQS(w)$. Furthermore, a $SQS(w)$ can be constructed which contains a $SQS(8)$ except if $w \in \{1, 2, 4, 10, 14\}$.*

Proof Trivially there exists a $SQS(w)$ for each $w \in \{1, 2, 4\}$. The cases $w = 8, 10$ and 14 are handled in Examples 7.1.3, 7.1.4 and 7.1.5 respectively. For each $w \in \{16, 20, 28, 32\}$, The $2v$ Construction produces a $SQS(w)$ containing a $SQS(8)$ (see Theorem 7.2.1). A $SQS(22)$ containing a $SQS(v = 8)$ can be formed using The $(3v - 2)$ Construction (see Exercise 7.2.7). If $w = 26, 34$ or 38 then a $SQS(w)$ containing a $SQS(8)$ is constructed in Examples 7.4.13, 7.4.14 (see Exercise 7.4.17) or 7.4.22 respectively.

The remaining values of w are handled inductively, so assume that for some $r \geq 40, r \equiv 2$ or 4 (mod 6), and that for all $16 \leq w < r$ there exists a $SQS(w)$ containing a $SQS(8)$ (notice that we have already constructed $SQS(w)$'s containing a $SQS(8)$ for all $w \equiv 2$ or 4 (mod 6), $16 \leq w \leq 38$). We now produce a $SQS(r)$ containing a $SQS(8)$. We consider several cases in turn.

Case 1: If $r \equiv 4$ or 8 (mod 12) then The $2v$ Construction produces a $SQS(r)$ containing a $SQS(8)$ (see Theorem 7.2.1).

Case 2: If $r \equiv 10$ or 22 (mod 36) then $r \geq 46$. The $(3v - 2)$ Construction with $v = (r + 2)/3 \geq 16$ produces a $SQS(r)$ containing a $SQS(v)$ (see Theorem 7.2.6) which by the inductive hypothesis contains a $SQS(8)$.

Case 3: If $r \equiv 14, 26$ or 34 (mod 36) then $r \geq 50$. The $(3v - 2u)$ Construction with $v \not\equiv u$ (mod 6) and $u = 8, 2$ or 4 produces a $SQS(r)$ for $r \equiv 14, 26$ or 34 (mod 36) respectively containing a $SQS(v)$ (see Lemma 7.4.12), which by the inductive hypothesis contains a $SQS(8)$ (in each case, $v \geq 16$).

Case 4: If $r \equiv 2$ (mod 36) then $r \geq 74$. The $(3v - 2u)$ Construction with $v \equiv u = 2$ and $v = (r + 2u)/3 \geq 26$ produces a $SQS(r)$ containing a $SQS(v)$ (see Lemma 7.4.21) which by the inductive hypothesis contains a $SQS(8)$.

This exhausts all possible congruence classes modulo 36, so the proof is complete. □

Example 7.4.27 We show how to construct a $SQS(230)$ using the proof of Theorem 7.4.26.
Since $230 \equiv 14$ (mod 36), using Case 3 we can let $v_1 = 82$, $u_1 = 8$ and apply The $(3v - 2u)$ Construction to a $SQS(82)$ that contains a $SQS(8)$ to produce a $SQS(3v - 2u = 230)$.

But how do we find the $SQS(82)$ containing a $SQS(8)$?

Since $82 \equiv 10$ (mod 36), using Case 2 we can let $v_2 = 28$ and use The $(3v-2)$ Construction to produce a $SQS(3v_2 - 2 = 82)$ containing a $SQS(v_3 = 28)$. So this will contain a $SQS(8)$ as required as long as we have a $SQS(28)$ containing a $SQS(8)$.

We can construct a $SQS(28)$ using The $2v$ Construction, starting with the $SQS(v_4 = 14)$ in Example 7.1.5. Since we use The $2v$ Construction, we can ensure that the resulting $SQS(28)$ contains a $SQS(8)$ (see Theorem 7.2.1).

Therefore, we have all the ingredients necessary to produce a $SQS(230)$.

Exercises

7.4.28 Use the proof of Theorem 7.4.26 to show how to construct a $SQS(v)$, for each of the following values of v.

(a) $v = 142$

(b) $v = 166$

(c) $v = 182$

(d) $v = 214$

(e) $v = 256$

Appendices

Appendix A

Cyclic Steiner Triple Systems

Rose Peltesohn [19] found the following difference triples. These can be used to form the base blocks (see Section 1.7) of a cylic $STS(v)$ for all $v \equiv 1$ or 3 (mod 6), $v \neq 9$. There does not exist a cyclic $STS(9)$.

v = 7 $\{1,2,3\}$
v = 13 $\{1,3,4\}$ and $\{2,5,6\}$
v = 15 $\{1,3,4\}$ and $\{2,6,7\}$
v = 19 $\{1,5,6\}$, $\{2,8,9\}$ and $\{3,4,7\}$
v = 27 $\{1,12,13\}$, $\{2,5,7\}$, $\{3,8,11\}$ and $\{4,6,10\}$
v = 45 $\{1,11,12\}$, $\{2,17,19\}$, $\{3,20,22\}$, $\{4,10,14\}$, $\{5,8,13\}$, $\{6,18,21\}$ and $\{7,9,16\}$
v = 63 $\{1,15,16\}$, $\{2,27,29\}$, $\{3,25,28\}$, $\{4,14,18\}$, $\{5,26,31\}$, $\{6,17,23\}$, $\{7,13,20\}$, $\{8,11,19\}$, $\{9,24,30\}$ and $\{10,12,22\}$.
v = 18s + 1, $s \geq 2$
$\{3r+1, 4s-r+1, 4s+2r+2\}$ for $0 \leq r \leq s-1$,
$\{3r+2, 8s-r, 8s+2r+2\}$ for $0 \leq r \leq s-1$,
$\{3r+3, 6s-2r-1, 6s+r+2\}$ for $0 \leq r \leq s-2$, and
$\{3s, 3s+1, 6s+1\}$.
v = 18s + 7, $s \geq 1$
$\{3r+1, 8s-r+3, 8s+2r+4\}$ for $0 \leq r \leq s-1$,
$\{3r+2, 6s-2r+1, 6s+r+3\}$ for $0 \leq r \leq s-1$,
$\{3r+3, 4s-r+1, 4s+2r+4\}$ for $0 \leq r \leq s-1$, and
$\{3s+1, 4s+2, 7s+3\}$.
v = 18s + 13, $s \geq 1$
$\{3r+2, 6s-2r+3, 6s+r+5\}$ for $0 \leq r \leq s-1$,
$\{3r+3, 8s-r+5, 8s+2r+8\}$ for $0 \leq r \leq s-1$,
$\{3r+1, 4s-r+3, 4s+2r+4\}$ for $0 \leq r \leq s$, and
$\{3s+2, 7s+5, 8s+6\}$.
v = 18s + 3, $s \geq 1$
$\{3r+1, 8s-r+1, 8s+2r+2\}$ for $0 \leq r \leq s-1$,
$\{3r+2, 4s-r, 4s+2r+2\}$ for $0 \leq r \leq s-1$, and
$\{3r+3, 6s-2r-1, 6s+r+2\}$ for $0 \leq r \leq s-1$.
v = 18s + 9, $s \geq 4$
$\{3r+1, 4s-r+3, 4s+2r+4\}$ for $0 \leq r \leq s$,

$\{3r+2, 8s-r+2, 8s+2r+4\}$ for $2 \leq r \leq s-2$,
$\{3r+3, 6s-2r+1, 6s+r+4\}$ for $1 \leq r \leq s-2$,
$\{2, 8s+3, 8s+5\}, \{3, 8s+1, 8s+4\}, \{5, 8s+2, 8s+7\}$,
$\{3s-1, 3s+2, 6s+1\}$, and $\{3s, 7s+3, 8s+6\}$.
v = 18s + 15 $s \geq 1$
$\{3r+1, 4s-r+3, 4s+2r+4\}$ for $0 \leq r \leq s$,
$\{3r+2, 8s-r+6, 8s+2r+8\}$ for $0 \leq r \leq s$, and
$\{3r+3, 6s-2r+3, 6s+r+6\}$ for $0 \leq r \leq s-1$.

Appendix B

Answers to Selected Exercises

Chapter 1

1.2.7

(a) v = 21

(b) (i) $\{(6, 1), (7, 2), (5, 1)\}$ (ii) $\{(5, 2), (5, 3), (5, 1)\}$
(iii) $\{(3, 1), (5, 1), (1, 2)\}$ (iv) $\{(3, 1), (5, 3), (4, 3)\}$
(v) $\{(4, 3), (7, 3), (1, 1)\}$ (vi) $\{(6, 1), (2, 2), (4, 1)\}$
(vii) $\{(3, 2), (7, 2), (2, 3)\}$

1.3.6

(a) v = 25

(b) (i) $\{(1, 1), (1, 3), (1, 2)\}$ (ii) $\{(6, 1), (6, 2), (7, 1)\}$
(iii) $\{(5, 1), \infty, (1, 2)\}$ (iv) $\{(4, 1), \infty, (8, 3)\}$
(v) $\{(4, 1), (6, 3), (3, 3)\}$ (vi) $\{(7, 1), (7, 3), (5, 3)\}$
(vii) $\{(4, 1), (5, 1), (3, 2)\}$ (viii) $\{(2, 3), (6, 2), \infty\}$
(ix) $\{(5, 2), (7, 2), (7, 3)\}$ (x) $\{(1, 3), (7, 1), (6, 3)\}$

1.4.2

$6n^2 + 9n$ triples.

1.4.5

(i) $\{\infty_1, (3, 2), (3, 3)\}$ (ii) $\{(3, 3), (5, 3), (1, 1)\}$

(iii) $\{(4, 3), (5, 1), \infty_1\}$ (iv) $\{(1, 1), (1, 3), (1, 2), \infty_1, \infty_2\}$

(v) $\{(3, 3), (4, 3), (2, 1)\}$ (vi) $\{(3, 1), (5, 3), (4, 3)\}$

(vii) $\{(2, 1), (5, 3), \infty_2\}$ (viii) $\{(1, 1), (5, 3), (3, 3)\}$

1.5.4

Many renamings are possible. For example, renaming 1, 2, 3, 4, 5, 6, 7, 8, 9, 10 and 11 with 1, 3, 5, 7, 10, 6, 4, 8, 9, 2 and 11 respectively produces the following blocks:

1 3 5 7 9	3 6 9	5 4 8	7 8 11
1 6 4	3 4 11	5 9 2	10 6 8
1 8 9	3 8 2	7 6 10	10 4 2
1 2 11	5 6 11	7 4 9	10 9 11

1.5.15

(a) $\{(2, 1), (3, 2), (5, 1)\}$ (b) $\{(3, 1), (4, 2), (4, 3)\}$

(c) $\{(5, 1), \infty, (6, 1)\}$ (d) $\{(5, 1), (5, 3), (5, 2)\}$

(e) $\{(4, 1), (5, 3), (7, 3)\}$ (f) $\{(7, 2), \infty, (8, 2)\}$

1.5.16

(a) $\{(2, 1), (4, 3), (8, 3)\}$ (b) $\{(3, 1), (4, 3), \infty_3\}$

(c) $\{(6, 2), (6, 3), (6, 1)\}$ (d) $\{(6, 2), \infty_3, (5, 2)\}$

(e) $\{(6, 2), (8, 1), (3, 1)\}$ (f) $\{(1, 1), \infty_2, (2, 1)\}$

1.5.17

(a) $\{(2, 1), \infty_2, (2, 3)\}$ (b) $\{(3, 1), (4, 1), \infty_1\}$

(c) $\{(3, 3), (5, 3), (2, 1)\}$ (d) $\{(1, 3), (2, 1), \infty_5\}$

(e) $\{(1, 3), \infty_4, (1, 1)\}$ (f) $\{\infty_2, \infty_4, \infty_1, \infty_3, \infty_5\}$

1.6.7

(a) $\{1, 27, \infty_1\}$ (b) $\{5, 51, 54\}$ (c) $\{7, 16, 32\}$ (d) $\{6, 7, 42\}$

(e) $\{2, \infty_1, 51\}$ (f) $\{2, 53, \infty_2\}$ (g) $\{5, 45, 0\}$ (h) $\{11, 14, 30\}$

1.7.3

(a) $\{1, 5, 6\}, \{2, 8, 9\}, \{3, 4, 7\}$

(b) $\{1, 9, 10\}, \{2, 4, 6\}, \{3, 5, 8\}$

(c) $\{1, 2, 3\}, \{4, 7, 11\}, \{5, 8, 12\}, \{6, 9, 10\}$

(d) $\{1, 2, 3\}, \{4, 10, 13\}, \{5, 6, 11\}, \{7, 8, 12\}$

(e) $\{1, 2, 3\}, \{4, 8, 12\}, \{5, 9, 14\}, \{6, 10, 15\}, \{7, 11, 13\}$

(f) $\{1, 15, 16\}, \{2, 4, 6\}, \{3, 7, 10\}, \{5, 9, 14\}, \{8, 12, 13\}$

1.7.8

(a) $\{0, 1, 6\}, \{0, 2, 10\}, \{0, 3, 7\}$

(b) $\{0, 1, 10\}, \{0, 2, 6\}, \{0, 3, 8\}$

(c) $\{0, 1, 3\}, \{0, 4, 11\}, \{0, 5, 13\}, \{0, 6, 15\}$

(d) $\{0, 1, 3\}, \{0, 4, 14\}, \{0, 5, 11\}, \{0, 7, 15\}$

(e) $\{0, 1, 3\}, \{0, 4, 12\}, \{0, 5, 14\}, \{0, 6, 16\}, \{0, 7, 18\}$

(f) $\{0, 1, 16\}, \{0, 2, 6\}, \{0, 3, 10\}, \{0, 5, 14\}, \{0, 8, 20\}$

1.7.9

(a) $\{7, 14, 5\}$ (b) $\{0, 2, 9\}$ (c) $\{7, 12, 2\}$

(d) $\{1, 12, 0\}$ (e) $\{0, 7, 13\}$ (f) $\{6, 10, 9\}$

2.3.9

(a) $\{\infty, (7,2), (7,1)\}, \{\infty, (7,2), (7,3)\}$

(b) $\{(2,2), (6,2), (1,3)\}, \{(2,2), (6,2), (7,3)\}$

(c) $\{(3,3), (4,1), (6,3)\}, \{(3,3), (4,1), (1,3)\}$

(d) $\{(2,1), (3,1), (7,2)\}, \{(2,1), (3,1), (5,2)\}$

(e) $\{(6,2), (5,1), (1,1)\}, \{(6,2), (5,1), (3,1)\}$

2.4.10

(a) $\{1,3,5\}, \{1,3,2\}, \{1,3,4\}$

(b) $\{2,5,1\}, \{2,5,4\}, \{2,5,3\}$

(c) $\{4,5,3\}, \{4,5,2\}, \{4,5,1\}$

2.4.11

(a) $\{1,4,5\}, \{1,5,2\}, \{3,5,1\}, \{5,1,3\}, \{5,4,1\}, \{2,5,1\}$

(b) $\{2,1,4\}, \{2,4,3\}, \{4,2,5\}, \{5,2,4\}, \{4,1,2\}, \{3,4,2\}$

(c) $\{3,5,4\}, \{3,1,5\}, \{5,3,2\}, \{2,3,5\}, \{5,1,3\}, \{4,5,3\}$

3.2.10

(i) leave

(ii) $\{\infty_1, (2,1), (2,2)\}$

(iii) leave

(iv) $\{(3,1), (3,3), (7,3)\}$

(v) $\{(1,1), (1,2), (1,3)\}$

(vi) $\{(3,2), (4,2), (6,3)\}$

3.3.8

(i) $\{\infty, \infty_2, (1,3)\}$

(ii) $\{\infty, (1,3), \infty_1\}, \{\infty, (1,3), \infty_2\}$

(iii) $\{\infty, (6,2), (7,2)\}$

(iv) $\{(4,1), (6,1), (2,2)\}$

(v) $\{(4,1), (5,1), (3,2)\}, \{(4,1), (5,1), \infty\}$

4.1.5

(a)

$$\begin{aligned}
B_6 &= \{\{3,4,6,12\},\{1,5,6,8\},\{2,6,7,9\},\{6,10,11,13\}\} \\
\pi_6 &= \{\{\infty,(6,1),(6,2)\},\{(3,1),(12,1),(4,2)\},\{(3,2),(4,1),\\
&\quad (12,2)\},\{(1,1),(8,1),(5,2)\},\{(1,2),(5,1),(8,2)\},\\
&\quad \{(2,1),(7,1),(9,2)\},\{(2,2),(9,1),(7,2)\},\\
&\quad \{(10,1),(11,1),(13,1)\},\{(10,2),(11,2),(13,2)\}\}
\end{aligned}$$

(d) (i) $\{(2, 1), (7, 2), (6, 1)\}$ (ii) $\{(2, 1), (8, 1), (12,2)\}$
(iii) $\{(3, 1), (3, 2), \infty\}$ (iv) $\{(5, 2), (6, 2), (8, 2)\}$
(v) $\{(5, 2), \infty, (5, 1)\}$

(e) (i) π_9 (ii) π_{13} (iii) π_3 (iv) π_1 (v) π_5

4.1.6

(a) (i) $(7,1)$ (ii) $(13,1)$ (iii) $(16,2)$ (iv) $(12,2)$

(b) (i) π_5 (ii) π_{20} (iii) π_{13} (iv) π_{15}

(c) (i) $\{(4,1),(13,2),(14,2)\}$ (ii) $\{(4,2),(19,1),(20,1)\}$
(iii) $\{(9,1),(16,1),(17,1)\}$ (iv) $\{(1,2),(2,1),(26,2)\}$
(v) $\{(3,1),(7,1),(8,2)\}$ (vi) $\{(10,2),(11,1),(23,2)\}$

4.2.2

$$\begin{aligned}\pi_{17} = & \{\{\infty,(17,1),(17,2)\},\{(1,1),(4,1),(5,2)\},\{(1,2),(5,1),(4,2)\},\\ & \{(2,1),(8,1),(10,2)\},\{(2,2),(10,1),(8,2)\}\\ & \{(3,1),(13,1),(15,2)\},\{(3,2),(15,1),(13,2)\}\\ & \{(6,1),(9,1),(14,2)\},\{(6,2),(14,1),(9,2)\}\\ & \{(7,1),(11,1),(12,2)\},\{(7,2),(12,1),(11,2)\},\\ & \{(16,1),(19,1),(20,1)\},\{(16,2),(21,2),(22,2)\}\\ & \{(18,1),(19,2),(22,1)\},\{(18,2),(20,2),(21,1)\}\}\end{aligned}$$

4.2.3 The following answers only count the blocks of sizes greater than 4.

v = 4 All blocks size 4 v = 7 One block of size 7
v = 10 One block of size 10 v = 13 All blocks have size 4
v = 16 All blocks have size 4 v = 19 One block of size 19
v = 22 One block of size 7 v = 25 All blocks have size 4
v = 28 Four blocks of size 7 v = 31 One block of size 10
v = 34 One block of size 7 v = 37 All blocks have size 4
v = 40 All blocks have size 4 v = 43 7 blocks of size 7
v = 46 5 blocks of size 10 v = 79 13 blocks of size 7
v = 82 3 blocks of size 7, 7 blocks of size 10, and 1 block of size 19

4.2.9

(a) m = 8 and t = 0: all blocks have size 4

(b) m = 8 and t = 2: one block of size 7, the rest of size 4

(c) m = 11 and t = 3: 4 blocks of size 7, 1 block of size 10, the rest of size 4

(d) m = 15 and t = 3: 21 blocks of size 10, the rest of size 4

5.1.3

1	2	3	4
3	4	1	2
4	3	2	1
2	1	4	3

1	2	3	4
4	3	2	1
2	1	4	3
3	4	1	2

1	2	3	4
2	1	4	3
3	4	1	2
4	3	2	1

5.2.9

(a) 0 (b) x^4 (c) x^{22} (d) x^6

5.2.10

(a) (x^{14}, x^{13}) (b) (x^{19}, x^{22}) (c) (x^7, x^{13})

5.2.19

(a) 3 (b) 4 (c) 80 (d) 6 (3) 4

7.1.11

Many renamings are possible. For example, renaming $\{1, 2, \}$, $\{1, 3\}$, $\{1, 4\}$, $\{1, 5\}$, $\{2, 3\}$, $\{2, 4\}$, $\{2, 5\}$, $\{3, 4\}$, $\{3, 5\}$, $\{4, 5\}$ with 7, 9, 1, 6, 8, 2, 5, 3, 4, 10 respectively means that the quadruples $\{1, 2, 3, 10\}$, $\{4, 5, 6, 10\}$, $\{7, 8, 9, 10\}$ and $\{3, 6, 9, 10\}$ correspond to copies of $K_{1,4}$, $K_{1,4}$, $K_3 + K_2$ and C_4 respectively.

7.1.18

$F_6 = \{\{1, 6\}, \{3, 5\}, \{4, 8\}, \{2, 7\}\}$.

7.2.4

(a) $\{(1, 1), (4, 1), (6, 1), (7, 1)\}$

(b) $\{(2, 2), (4, 2), (5, 2), (7, 2)\}$

(c) $\{(1, 1), (6, 1), (7, 2), (2, 2)\}$

(d) $\{(2, 2), (6, 2), (3, 1), (8, 1)\}$

(e) Does not exist.

7.2.7

(i) $\{\infty, (2, 1), (4, 3), (6, 2)\}$

(ii) $\{(2, 1), (2, 3), (7, 2), (7, 3)\}$

(iii) $\{(3, 1), (4, 1), (5, 1), (6, 3)\}$

(iv) $\{(1, 1), (2, 1), (7, 1), \infty\}$

(v) $\{(2,1),(3,3),(5,1),(5,3)\}$

7.4.3

(a) $T_3(10)$ (f) T_1

(b) $T_2(3)$ (g) T_1

(c) T_1 (h) T_1

(d) $T_4(4)$ (i) $T_3(5)$

(e) $T_4(6)$ (j) $T_4(2)$

7.4.9

(a) $\{(2,3),(3,3),(6,3),(5,3)\}$

(b) $\{(2,2),(5,2),\infty_1,(8,2)\}$

(c) $\{(1,1),(6,3),\infty_1,(7,2)\}$

(d) $\{(1,1),(7,2),(6,3),(7,3)\}$

(e) $\{(2,1),(7,2),(6,3),(4,2)\}$

(f) $\{(1,1),(4,1),(6,3),(3,2)\}$

(g) $\{(1,2),(8,2),(6,3),(6,1)\}$

7.4.15

(a) $3v - 2u = 26$; $v = 10$; $u = 2$; $g = v - u = 8$.
For example, use: $Q_1, Q_2(2,1), Q_2(3,2), Q_3(4,6,0;1), Q_4(\{2,3,4\})$,
(b) $3v - 2u = 34, v = 14, u = 4, v - u = g = 10$
For example, use: $Q_1, Q_2(0,1), Q_2(1,2), Q_2(2,3), Q_2(3,4), Q_3(4,6,8;1)$, $Q_4(\{2,3,4,5\})$.

Bibliography

[1] R. C. Bose, *On the construction of balanced incomplete block designs*, Ann. Eugenics, **9** (1939), 353-399.

[2] R. C. Bose and S. S. Shrikhande, On the falsity of Euler's conjecture about the nonexistence of two orthogonal latin squares of order $4t+2$, Proc. Nat. Acad. Sci. USA, **45** (1959), 734-737.

[3] R. C. Bose, S. S. Shrikhande and E. T. Parker, *Further results on the construction of mutually orthogonal latin squares and the falsity of Euler's conjecture*, Canadian J. Math., **12** (1960), 189-203.

[4] R. H. Bruck and H. J. Ryser, *The nonexistence of certain finite projective planes*, Canadian J. Math., **1** (1949), 88-93.

[5] I. Diener, E. Schmitt and H. L. deVries, All 80 Steiner triple systems on 15 elements are derived, Discrete Math., **55** (1985), 13-19.

[6] A. L. Dulmage, D. Johnson and N. S. Mendelsohn, *Orthomorphisms of groups and orthogonal Latin squares*, Canadian J. Math., **13** (1961), 356-372.

[7] L. Euler, Recherches sur une nouvelie espèce de quarres magiques, Vehandlingen Zeeuwach Genootschap Wetenshapper Vlissengen, **9** (1782), 85-239.

[8] H. Hanani, *On quadruple systems*, Canadian J. Math., **12** (1960), 145-157.

[9] A. Hartman, *Tripling quadruple systems*, Ars Combinatoria, **10** (1980), 255-309.

[10] A. Hartman, *A general recursion construction for quadruple systems*, J. Comb. Theory (A), **33** (1982), 121-134.

[11] L. Heffter, *Über Nachbarconfigurationen, Tripelsysteme und metacyklische Gruppen*, Deutsche Mathem. Vereining. Jahresber., **5** (1896), 67-69.

[12] T. P. Kirkman, *On a problem in combinations*, Cambridge and Dublin Math. Journal, **2** (1847), 191-204.

[13] T. P. Kirkman, *Query VI.*, Lady's and Gentleman's Diary, (1850), 48.

[14] T. P. Kirkman, *Solution to Query VI.*, Lady's and Gentleman's Diary, (1851), 48.

[15] H. Lenz, *Tripling Steiner Quadruple Systems*, Ars. Combinatoria, **20** (1985), 193-202.

[16] H. F. MacNeish, *Euler squares*, Ann. Math., **23** (1922), 221-227.

[17] E. T. Parker, *Orthogonal latin squares*, Proc. Nat. Acad. Sci. U.S.A., **45** (1959), 859-862.

[18] E. T. Parker, *Construction of some sets of mutually orthogonal latin squares*, Proc. Amer. Math. Soc., (1959), 946-949.

[19] R. Peltesohn, *Eine Lösung der beiden Heffterschen Differenzenprobleme*, Compositio Math., **6** (1939), 251-257.

[20] D. K. Ray-Chaudhuri and R. M. Wilson, *Solution of Kirkman's school-girl problem*, Proc. Symp. Pure Math., Amer. Math. Soc., Providence, RI, **19** (1971), 187-203.

[21] Th. Skolem, *Some remarks on the triple systems of Steiner*, Math. Scand., **6** (1958), 273-280.

[22] G. Stern and H. Lenz, *Steiner triple systems with given sub-spaces; another proof of the Doyen-Wilson theorem*, Boll. Un. Mat. Ital. (5), 17 - A (1980), 109-114.

[23] D. R. Stinson, A short proof of the nonexistence of a pair of orthogonal Latin squares of order six, J. Combinatorial Th. (A), **36** (1984), 373-376.

[24] G. Tarry, Le problème des 36 officers, C. R. Assoc. Fr. Av. Sci. **29** (1900), 170-203.

[25] V. G. Vizing, On an estimate of the chromatic class of a p-graph, Diskret Analiz, **3** (1964), 25-30.

[26] R. M. Wilson, An existence theory for pairwise balanced designs: 1. Composition theorems and morphisms, J. Combinatorial Th., **13** (1972), 220-244.

[27] R. M. Wilson, *Some partitions of all triples into Steiner triple systems*, Lecture Notes in Math. 411, Springer, Berlin, (1974), 267-277.

[28] W. S. B. Woolhouse, *Prize question 1733*, Lady's and Gentleman's Diary, (1844).

Index